Matthew M. Muir

Heroes of Science

Chemists

Matthew M. Muir

Heroes of Science
Chemists

ISBN/EAN: 9783337034627

Printed in Europe, USA, Canada, Australia, Japan

Cover: Foto ©berggeist007 / pixelio.de

More available books at **www.hansebooks.com**

HEROES OF SCIENCE.

CHEMISTS.

BY

M. M. PATTISON MUIR, M.A., F.R.S.E.,

FELLOW, AND PRÆLECTOR IN CHEMISTRY, OF GONVILLE
AND CAIUS COLLEGE, CAMBRIDGE.

PUBLISHED UNDER THE DIRECTION OF THE COMMITTEE
OF GENERAL LITERATURE AND EDUCATION APPOINTED BY THE
SOCIETY FOR PROMOTING CHRISTIAN KNOWLEDGE.

LONDON:
SOCIETY FOR PROMOTING CHRISTIAN KNOWLEDGE,
NORTHUMBERLAND AVENUE, CHARING CROSS;
43, QUEEN VICTORIA STREET, E.C. ;
26, ST. GEORGE'S PLACE, HYDE PARK CORNER, S.W.
BRIGHTON: 135, NORTH STREET.
NEW YORK: E. & J. B. YOUNG & CO.
1883.

"The discoveries of great men never leave us; they are immortal; they contain those eternal truths which survive the shock of empires, outlive the struggles of rival creeds, and witness the decay of successive religions."—BUCKLE.

"He who studies Nature has continually the exquisite pleasure of discerning or half discerning and divining laws; regularities glimmer through an appearance of confusion, analogies between phenomena of a different order suggest themselves and set the imagination in motion; the mind is haunted with the sense of a vast unity not yet discoverable or nameable. There is food for contemplation which never runs short; you gaze at an object which is always growing clearer, and yet always, in the very act of growing clearer, presenting new mysteries."—THE AUTHOR OF "ECCE HOMO."

"Je länger ich lebe, desto mehr verlern' ich das Gelernte, nämlich die Systeme."—JEAN PAUL RICHTER.

PREFACE.

I HAVE endeavoured in this book to keep to the lines laid down for me by the Publication Committee of the Society, viz. "to exhibit, by selected biographies, the progress of chemistry from the beginning of the inductive method until the present time." The progress of chemistry has been made the central theme; around this I have tried to group short accounts of the lives of those who have most assisted this progress by their labours.

This method of treatment, if properly conducted, exhibits the advances made in science as intimately connected with the lives and characters of those who studied it, and also impresses on the reader the continuity of the progress of natural knowledge.

The lives of a few chemists have been written; of others there are, however, only scanty notices to be found. The materials for this book have been collected chiefly from the following works :—

Kopp's "Geschichte der Chemie."
Thomson's "History of Chemistry."
Ladenburg's "Entwickelungsgeschichte der Chemie."
Wurtz's "History of the Atomic Theory."
Watts's "Dictionary of Chemistry."
Whewell's "History of the Inductive Sciences."
Rodwell's "Birth of Chemistry;" "Inquiry into the Hermetic Mystery and Alchemy" (London, 1850); "Popular Treatises on Science written during the Middle Ages," edited for the Historical Society of Science by Thomas Wright, M.A. (London, 1841); "Ripley Reviv'd; or, An Exposition upon Sir George Ripley's Hermetico-Poetical Works," by Eirenæus Philalethes (London, 1678); "Tripus Aureus, hoc est Tres Tractates Chymici Selectissimi" (Frankfurt, 1618).
"Alchemy;" article in "Encyclopædia Britannica."
Boyle's "Sceptical Chymist."
"Biographie Universelle;" for notices of Berzelius and Lavoisier.
"English Cyclopædia;" for notices of Black, Berzelius and Lavoisier.
Black's "Lectures," with Memoir: edited by Dr. Robinson.
Priestley's "Memoirs:" written partly by himself.
Priestley's works on "Air," etc.
Lavoisier's "Œuvres."
Dalton's "Life," by Dr. Henry; "Life," by Dr. R. Angus Smith; "New System of Chemical Philosophy."
Davy's "Collected Works;" with Life, by his brother; "Life," by Dr. Paris.

Berzelius's " Lehrbuch," and various dissertations.
Wöhler's " Jugenderinnerungen eines Chemikers."
Graham's " Collected Memoirs."
Sketch of Graham's life, in Chemical Society's Journal.
" Life-Work of Liebig," by A. W. Hofmann.
" Dumas," by A. W. Hofmann.
Various dissertations by Liebig and Dumas in *Annalen*,
and elsewhere.

My warmest thanks are due to my friend, Mr.
Francis Rye, for the great assistance he has given
me in correcting the proof-sheets.

M. M. PATTISON MUIR.

Cambridge, *April*, 1883.

CONTENTS.

HEROES OF SCIENCE.

As we trace the development of any branch of natural knowledge we find that there has been a gradual progress from vague and fanciful to accurate and definite views of Nature. We find that as man's conceptions of natural phenomena become more accurate they also for a time become more limited, but that this limitation is necessary in order that facts may be correctly classified, and so there may be laid the basis for generalizations which, being definite, shall also be capable of expansion.

At first Nature is strange ; she is full of wonderful and fearful appearances. Man is overwhelmed by the sudden and apparently irregular outbreaks of storms, by the capricious freaks of thunder and lightning, by the awful and unannounced devastations of the volcano or the earthquake ; he believes himself to be surrounded by an invisible array of

III.   B

beings more powerful than himself, but, like himself, changeable in their moods and easily provoked to anger. After a time he begins to find that it is possible to trace points of connection between some of the appearances which had so overpowered or perplexed him.

The huntsman observes that certain kinds of plants always grow where the game which he pursues is chiefly to be found ; from the appearance of the sky at morning and evening the fisherman is able to tell whether there will follow weather suitable for him to set out in his fishing-boat ; the tiller of the ground begins to feel sure that if he sow the seed in the well-dug soil and water it in proper seasons he will certainly reap the harvest in due time. And thus man comes to believe that natural events follow each other in a fixed order ; there arises a conscious reference on his part of certain effects to certain definite causes. Accurate knowledge has begun.

As knowledge of natural appearances advances there comes a time when men devote themselves chiefly to a careful study of some one class of facts ; they try to consider that part of Nature with which they are mostly concerned as separate from all other parts of Nature. Thus the various branches of natural knowledge begin to have each a distinct existence. These branches get more and more subdivided, each division is more accurately studied, and so a great number of facts is accumulated in many classes. Then we usually find that a master mind arises, who shows the connection which exists between the different parts of each division of natural knowledge, who takes a wide, far-reaching view of the whole range of the province of knowledge which he studies, and who, at the same time, is able to hold in his vision all the important details of each branch of which that province is composed.

And thus we again get wide views of Nature. But these are very different from the vague, dim and hesitating notions in which natural knowledge had its beginnings. In this later time men see that Nature is both simple and complex; that she is more wonderful than their fathers dreamed, but that through all the complexity there runs a definite purpose ; that the apparently separate facts are bound together by definite laws, and that to discover this purpose and these laws is possible for man.

As we trace this progress in the various branches of natural knowledge we are struck with the fact that each important advance is generally accomplished by one or two leading men ; we find that it becomes possible to group the history of each period round a few central figures ; and we also learn that the character of the work done by each of these men of note is dependent on the nature and training of the individual man.

It will be my endeavour in the following pages to give an account of the advance of chemical science, grouping the facts in each stage of progress round the figures of one or two men who were prominent in that period.

For the purposes of this book it will be necessary that I should sketch only the most important periods in the story of chemical progress, and that in each of these I should fill in the prominent points alone.

I shall therefore select three periods in the progress of this science, and try to give an account of the main work done in each of these. And the periods will be :—

I. The period wherein, chiefly by the work of Black, Priestley and Lavoisier, the aim of chemical science was defined and the essential characters of the phenomena to be studied were clearly stated.

II. The period during which, chiefly by the labours of Dalton, Berzelius and Davy, the great central propositions of the science were laid down and were developed into a definite theory. As belonging in great extent to this period, although chronologically later, I shall also consider the work of Graham.

III. The period when, chiefly owing to advances made in organic chemistry, broader and more far-reaching systems of classification were introduced, and the propositions laid down in the preceding period were modified and strengthened. The workers in this period were very numerous ; I shall chiefly consider these two—Liebig and Dumas.

I shall conclude with a brief sketch of some of the important advances of chemical science in more recent times, and a summary of the characteristics of each of the three periods.

CHAPTER I.

ALCHEMY: AND THE DAWN OF CHEMISTRY.

EARLY chemistry was not a science. The ancient chemists dealt chiefly with what we should now call chemical manufactures; they made glass, cleaned leather, dyed cloth purple and other colours, extracted metals from their ores, and made alloys of metals. No well-founded explanations of these processes could be expected either from men who simply used the recipes of their predecessors, or from philosophers who studied natural science, not by the help of accurate experiments, but by the unaided light of their own minds.

At somewhat later times chemistry assumed a very important place in the general schemes propounded by philosophers.

Change is vividly impressed on all man's surroundings: the endeavour to find some resting-place amidst the chaos of circumstances, some unchanging substance beneath the ever-changing appearances of things, has always held a prominent place with those who study the phenomena of the world which surrounds them. In the third and fourth centuries of our era much attention was given to the art which professed to explain the changes of Nature. Religion, philosophy, and what we should now call natural science, were at that time closely intermingled ; the scheme of things which then, and for several centuries after that time, exerted a powerful influence over the minds of many thinkers was largely based on the conception of a fundamental unity underlying and regulating the observed dissimilarities of the universe.

Thus, in the *Emerald Table of Hermes*, which was held in much repute in the Middle Ages, we read—

" True, without error, certain and most true : that which is above is as that which is below, and that which is below is as that which is above, for performing the miracles of the *One Thing ;* and as all things were from one, by the mediation of one, so all things arose from this one thing by adaptation : the father of it is the Sun, the mother of it is the Moon, the wind carried it in its belly, the nurse of it is the Earth. This is the father of all perfection, the consummation of the whole world."

And again, in a later writing we have laid down the basis of the art of alchemy in the proposition that " there abides in nature a certain pure matter, which, being discovered and brought by art to perfection, converts to itself proportionally all imperfect bodies that it touches."

To discover this fundamental principle, this *One Thing*, became the object of all research. Earth and the heavens were supposed to be bound together by the all-pervading presence of the One Thing; he who should attain to a knowledge of this precious essence would possess all wisdom. To the vision of those who pursued the quest for the One Thing the whole universe was filled by one ever-working spirit, concealed now by this, now by that veil of sense, ever escaping identification in any concrete form, yet certainly capable of being apprehended by the diligent searcher.

Analogy was the chief guide in this search. If it were granted that all natural appearances were manifestations of the activity of one essential principle, then the vaguest and most far-fetched analogies between the phenomena of nature might, if properly followed up, lead to the apprehension of this hidden but everywhere present essence.

The history of alchemy teaches, in the most striking manner, the dangers which beset this method of pursuing the study of Nature; this history teaches us that analogies, unless founded on carefully and accurately determined facts, are generally utterly misleading in natural science.

Let us consider the nature of the experimental evidence which an alchemist of the fourth or fifth century could produce in favour of his statement that transmutation of one kind of matter into another is of constant occurrence in Nature.

The alchemist heated a quantity of water in an open glass vessel; the water slowly disappeared, and when it was all gone there remained in the vessel a small quantity of a white earthy solid substance. What could this experiment teach save that water was changed into earth and air? The alchemist then plunged a piece of red-hot iron into water placed under a bell-shaped glass vessel; some of the water seemed to be changed into air, and a candle, when brought into the bell, caused the air therein to take fire. Therefore, concluded the experimenter, water is proved to be changeable into fire.

A piece of lead was then strongly heated in the air; it lost its lustre and became changed into a reddish-white powder, very unlike lead in its properties; this powder was then heated in a convenient vessel with a little wheat, whereupon the lead was again produced. Therefore, said the alchemist, lead is destroyed by fire, but it can be reproduced from its ashes by the help of heat and a few grains of corn.

The experimenter would now proceed to heat a quantity of a mineral containing lead in an open vessel made of pulverized bones; the lead slowly disappeared, and at the close of the experiment a button of silver remained. Might he not triumphantly assert that he had transmuted lead into silver?

In order that the doctrine of the transmutation of metals might rest on yet surer evidence, the alchemist placed a piece of copper in spirits of nitre (nitric acid); the metal disappeared; into the green liquid thus produced he then placed a piece of iron; the copper again made its appearance, while the iron was removed. He might now well say that if lead was thus demonstrably changed into silver, and copper into iron, it was, to say the least, extremely probable that any metal might be changed into any other provided the proper means for producing the change could be discovered.

But the experimental alchemist had a yet stranger transmutation wherewith to convince the most sceptical. He poured mercury in a fine stream on to melted sulphur; at once the mercury and the sulphur disappeared, and in their place was found a solid substance black as the raven's wing. He then heated this black substance in a closed vessel, when it also disappeared, and in its place there was found, deposited on the cooler part of the vessel, a brilliantly red-coloured solid. This experiment taught lessons alike to the alchemist, the philosopher, and the moralist of these times. The alchemist learned that to change one kind of matter into another was an easy task: the philosopher learned that the prevalence of change or transmutation is one of the laws of Nature: and the moralist learned that evil is not wholly evil, but contains also some germs of good; for was not the raven-black substance emblematical of the evil,

and the red-coloured matter of the good principle
of things ? *

On such experimental evidence as this the
building of alchemy was reared. A close relationship was believed to prevail through the whole
phenomena of Nature. What more natural then
than to regard the changes which occur among the
forms of matter on this earth as intimately connected with the changes which occur among the
heavenly bodies ?

Man has ever been overawed by the majesty of
the stars ; yet he has not failed to notice that the
movements of these bodies are apparently capricious. The moon has always been to him a type
of mutability ; only in the sun has he seemed to
find a settled resting-point. Now, when we remember that in the alchemical scheme of things
the material earth and material heavens, the intellectual, the moral, and the spiritual world were
regarded as one great whole, the parts of which
were continuously acting and reacting on each
other, we cannot wonder that the alchemist should
regard special phenomena which he observed in
his laboratory, or special forms of matter which he
examined, as being more directly than other phenomena or other forms of matter, under the influence
of the heavenly bodies. This connection became
gradually more apparent to the student of alchemy,

* I have borrowed these illustrations of the alchemical experimental method from M. Hoefer's " Histoire de la Chimie," quoted
in the " Encyclopædia Brittanica," art. " Alchemy."

until at last it was fixed in the language and the symbols which he employed.

Thus the sun (Sol) was represented by a circle, which likewise became the symbol for gold, as being the most perfect metal. The moon (Luna) was ever changing ; she was represented by a half-circle, which also symbolized the pale metal silver.

Copper and iron were regarded as belonging to the same class of metals as gold, but their less perfect nature was denoted by the sign + or ↑. Tin and lead belonged to the lunar class, but like copper they were supposed to be imperfect metals. Mercury was at once solar and lunar in its properties.

These suppositions were summed up in such alchemical symbols as are represented below—

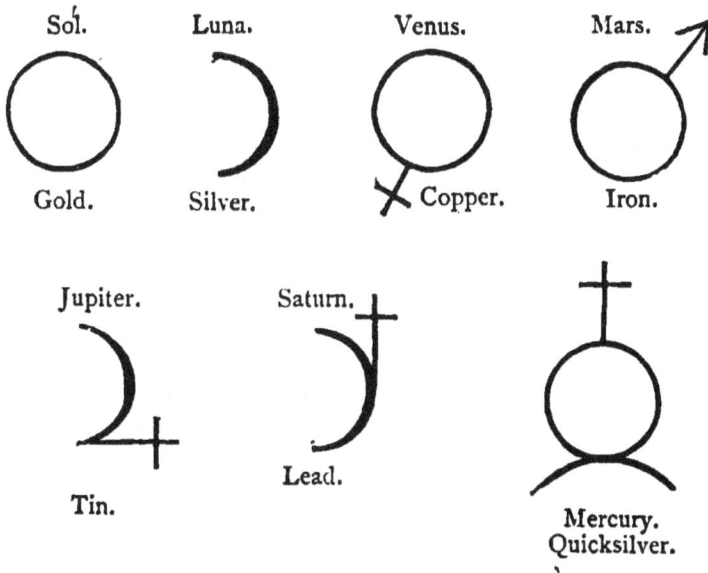

Sol.   Luna.   Venus.   Mars.

Gold.   Silver.   Copper.   Iron.

Jupiter.   Saturn.

Tin.   Lead.

Mercury.
Quicksilver.

Many of the alchemical names remain to the present time; thus in pharmacy the name "lunar caustic" is applied to silver nitrate, and the symptoms indicative of lead-poisoning are grouped together under the designation of "saturnine cholic."

But as the times advanced the older and nobler conception of alchemy became degraded.

If it be true, the later alchemists urged, that all things suffer change, but that a changeless essence or principle underlies all changing things, and that the presence of more or less of this essence confers on each form of matter its special properties, it follows that he who can possess himself of this principle will be able to transmute any metal into any other; he will be able to change any metal into gold.

Now, as the possession of gold has always carried with it the means of living luxuriously, it is easy to understand how, when this practical aspect of alchemy had taken firm root in men's minds, the pursuit of the art became for all, except a few lofty and noble spirits, synonymous with the pursuit of wealth. So that we shall not, I think, much err if we describe the chemistry of the later Middle Ages as an effort to accumulate facts on which might be founded the art of making gold. In one respect this was an advance. In the early days of alchemy there had been too much trusting to the mental powers for the manufacture of natural facts: chemists now actually worked in laboratories ; and very hard did many of these alchemists work.

Paracelsus says of the alchemists, " They are not given to idleness, nor go in a proud habit, or plush and velvet garments, often showing their rings upon their fingers, or wearing swords with silver hilts by their sides, or fine and gay gloves upon their hands ; but diligently follow their labours, sweating whole days and nights by their furnaces. They do not spend their time abroad for recreation, but take delight in their laboratory. They put their fingers amongst coals, into clay and filth, not into gold rings. They are sooty and black like smiths and miners, and do not pride themselves upon clean and beautiful faces." By thus " taking delight in their laboratories " the later alchemists gathered together many facts ; but their work centred round one idea, viz. that metals might all be changed into gold, and this idea was the result rather of intellectual guessing than of reasoning on established facts of Nature.

One of the most famous alchemists of the Middle Ages was born at Einsiedeln, in Switzerland, in 1493. His name, when paraphrased into Greek, became Paracelsus. This man, some of whose remarks have just been quoted, acquired great fame as a medical practitioner, and also as a lecturer on medicine : he travelled throughout the greater part of Europe, and is supposed to have been taught the use of several new medicines by the Arabian physicians whom he met in Spain. With an over-

weening sense of his own powers, with an ardent
and intemperate disposition, revolting against all
authority in medicine or science, Paracelsus yet
did a good work in calling men to the study of
Nature as the only means whereby natural science
could be advanced.

"Alchemy has but one aim and object," Paracelsus taught : "to extract the quintessence of
things, and to prepare arcana and elixirs which
may serve to restore to man the health and sound-
ness he has lost." He taught that the visible
universe is but an outer shell or covering, that
there is a spirit ever at work underneath this veil
of phenomena ; but that all is not active : "to
separate the active function (the spirit) of this
outside shell from the passive" was, he said, the
proper province of alchemy.

Paracelsus strongly insisted on the importance
of the changes which occur when a substance
burns, and in doing this he prepared the way for
Stahl and the phlogistic chemists.

However we may admire the general conceptions
underlying the work of the earlier alchemists, we
must admit that the method of study which they
adopted could lead to very few results of lasting
value ; and I think we may add that, however
humble the speculations of these older thinkers
might appear, this humility was for the most part
only apparent.

These men were encompassed (as we are) by unexplained appearances : they were every moment

reminded that man is not "the measure of all things;" and by not peering too anxiously into the mysteries around them, by drawing vague conclusions from partially examined appearances, they seemed at once to admit their own powerlessness and the greatness of Nature. But I think we shall find, as we proceed with our story, that this is not the true kind of reverence, and that he is the really humble student of Nature who refuses to overlook any fact, however small, because he feels the tremendous significance of every part of the world of wonders which it is his business and his happiness to explore.

As examples of the kind of explanation given by alchemists of those aspects of Nature which they professed to study, I give two quotations from translations of the writings of Basil Valentine and Paracelsus, who flourished in the first half of the fifteenth and sixteenth centuries respectively.

"Think most diligently about this; often bear in mind, observe and comprehend that all minerals and metals together, in the same time, and after the same fashion, and of one and the same principal matter, are produced and generated. That matter is no other than a mere vapour, which is extracted from the elementary earth by the superior stars, or by a sidereal distillation of the macrocosm; which sidereal hot infusion, with an airy sulphureous property, descending upon inferiors, so acts and operates as that there is implanted, spiritually and invisibly, a certain power and virtue

in those metals and minerals; which fume, moreover, resolves in the earth into a certain water
wherefrom all metals are thenceforth generated
and ripened to their perfection, and thence proceeds this or that metal or mineral, according as
one of the three principles acquires dominion and
they have much or little of sulphur and salt, or
an unequal mixture of these; whence some metals
are fixed, that is, constant or stable; and some
are volatile and easily changeable, as is seen in
gold, silver, copper, iron, tin and lead."

"The life of metals is a secret fatness; of salts,
the spirit of aqua fortis; of pearls, their splendour;
of marcasites and antimony, a tingeing metalline
spirit; of arsenics, a mineral and coagulated
poison. The life of all men is nothing else but
an astral balsam, a balsamic impression, and a
celestial invisible fire, an included air, and a tingeing spirit of salt. I cannot name it more plainly,
although it is set out by many names."

When the alchemists gave directions for making
the stone which was to turn all it touched into
gold, they couched them in such strange and
symbolical language as this: "After our serpent
has been bound by her chain, penetrated with the
blood of our green dragon, and driven nine or ten
times through the combustible fire into the elementary air, if you do not find her to be exceeding
furious and extremely penetrating, it is a sign that
you do not hit our subject, the notion of the
homogenea, or their proportion; if this furious

serpent does not come over in a cloud and turn into our virgin milk, or argentine water, not corrosive at all and yet insensibly and invisibly devouring everything that comes near it, it is plainly to be seen that you err in the notion of our universal menstruum." Or, again, what could any reasonable man make of this? "In the green lion's bed the sun and moon are born; they are married and beget a king. The king feeds on the lion's blood, which is the king's father and mother, who are at the same time his brother and sister. I fear I betray the secret, which I promised my master to conceal in dark speech from any one who knows not how to rule the philosopher's fire."

Concerning the same lion, another learned author says that "though called a lion, it is not an animal substance, but for its transcendant force, and the rawness of its origin, it is called the green lion." But he adds in a moment of confidence: "This horrid beast has so many names, that unless God direct the searcher it is impossible to distinguish him."

And once more. "Take our two serpents, which are to be found everywhere on the face of the earth: tie them in a love-knot and shut them up in the Arabian *caraha*. This is the first labour; but the next is more difficult. Thou must encamp against them with the fire of nature, and be sure thou dost bring thy line round about. Circle them in and stop all avenues that they find no relief.

III.   C

Continue this siege patiently, and they turn into
an ugly venomous black toad, which will be trans-
formed to a horrible devouring dragon, creeping
and weltering in the bottom of her cave without
wings.   Touch her not by any means, for there is
not on earth such a vehement transcending poison.
As thou hast begun so proceed, and this dragon
will turn into a swan.   Henceforth I will show
thee how to fortify thy fire till the phœnix appear :
it is a red bird of a most deep colour, with a shining
fiery hue.   Feed this bird with the fire of his father
and the ether of his mother : for the first is meat
and the second is drink, and without this last he
attains not to his full glory.   Be sure to understand
this secret," etc., etc.

The alchemists spoke of twelve gates through
which he who would attain to the palace of true
art must pass : these twelve gates were to be un-
locked by twelve keys, descriptions of which,
couched in strange and symbolical language, were
given in alchemical treatises.   Thus in " Ripley
reviv'd "* we read that Canon Ripley, of Bridling-
ton, who lived in the time of Edward IV., sang
thus of the first gate, which was " Calcination : "—

" The battle's fought, the conquest won,
    The Lyon dead reviv'd ;
The eagle's dead which did him slay,
    And both of sense depriv'd.

---

* " Ripley reviv'd : or an exposition upon Sir George Ripley's
Hermetico-poetical works," by Eirenæus Philalethes. London, 1678.

> The showers cease, the dews which fell
> For six weeks do not rise ;
> The ugly toad that did so swell
> With swelling bursts and dies."

And of the third gate, or "Conjunction," we find the Canon saying—

> " He was a king, yet dead as dead could be ;
> His sister a queen,
> Who when her brother she did breathless see,
> The like was never seen,
> She cryes
> Until her eyes
> With over-weeping were waxed dim—
> So long till her tears
> Reach'd up to her ears :
> The queen sunk, but the king did swim."

In some books these gates and keys are symbolically represented in drawings, *e.g.* in a pamphlet by Paracelsus, called "Tripus Aureus, hoc est Tres Tractates chymici selectissimi." (Frankfurt, 1618.)

It is evident that a method of studying Nature which resulted in such dim and hazy explanations as these was eminently fitted to produce many who pretended to possess secrets by the use of which they could bring about startling results beyond the power of ordinary men ; and, at the same time, the almost universal acceptance of such statements as those I have quoted implied the existence in men generally of a wondrous readiness to believe anything and everything. Granted that a man

by "sweating whole days and nights by his furnaces" can acquire knowledge which gives him
great power over his fellows, it necessarily follows
that many will be found ready to undergo these
days and nights of toil. And when we find that
this supposed knowledge is hidden under a mask
of strange and mystical signs and language, we
may confidently assert that there will be many
who learn to repeat these strange terms and use
these mystical signs without attempting to penetrate to the truths which lie behind—without, indeed, believing that the mystical machinery which
they use has any real meaning at all.

We find, as a matter of fact, that the age of
the alchemists produced many deceivers, who, by
mumbling incantations and performing a few
tricks, which any common conjuror would now
despise, were able to make crowds of men believe
that they possessed a supernatural power to con-
trol natural actions, and, under this belief, to make
them part with their money and their substance.

One respectable physician of the Hague, who
entertained a peripatetic alchemist, complains that
the man entered his "best-furnished room without
wiping his shoes, although they were full of snow
and dirt." However, the physician was rewarded,
as the stranger gave him, "out of his philosophical
commiseration, as much as a turnip seed in size"
of the much-wished-for stone of wisdom.

That the alchemist of popular belief was a man
who used a jargon of strange and high-sounding

words, that he might the better deceive those whom he pretended to help, is evident from the literature of the sixteenth and seventeenth centuries.

In the play of the "Alchymist" Ben Jonson draws the character of Subtle as that of a complete scoundrel, whose aim is to get money from the pockets of those who are stupid enough to trust him, and who never hesitates to use the basest means for this end. From the speeches of Subtle we may learn the kind of jargon employed by the men who pretended that they could cure diseases and change all baser metals into gold.

"*Subtle.* Name the vexations and the martyrizations of metals in the work.
Face. Sir, putrefaction,
Solution, ablution, sublimation,
Cohobation, calcination, ceration, and
Fixation.
Sub. And when comes vivification?
Face. After mortification.
Sub. What's cohobation?
Face. 'Tis the pouring on
Your aqua regis, and then drawing him off,
To the trine circle of the seven spheres.

    *   *   *   *   *   *

Sub. And what's your mercury?
Face. A very fugitive; he will be gone, sir.
Sub. How know you him?
Pace. By his viscosity,
His oleosity, and his suscitability."

Even in the fourteenth century, Chaucer (in the "Canon's Yeoman's Tale") depicts the alchemist as a mere cunning knave. A priest is prevailed

on to give the alchemist money, and is told that
he will be shown the change of base metal into
gold.   The alchemist busies himself with prepara-
tions, and sends the priest to fetch coals.

> "And whil he besy was, this feendly wrecche,
> This false chanoun (the foule feende him fecche)
> Out of his bosom took a bechen cole
> In which ful subtilly was maad an hole,
> And therein put was of silver lymayle
> An unce, and stopped was withoute fayle
> The hole with wex, to keep the lymayle in.
> And understondith, that this false gyn
> Was not maad there, but it was maad before."

This "false gyn" having been put in the crucible
and burned with the rest of the ingredients, duly
let out its "silver lymayle" (filings), which appeared
in the shape of a small button of silver, and so
accomplished the "false chanoun's" end of de-
ceiving his victim.

The alchemists accumulated many facts: they
gained not a little knowledge concerning the
appearances of Nature, but they were dominated
by a single idea.   Living in the midst of an extremely complex order of things, surrounded by a
strange and apparently capricious succession of
phenomena, they were convinced that the human
intelligence, directed and aided by the teachings
of the Church, would guide them through the
labyrinth.   And so they entered on the study of
Nature with preconceived notions and foregone
conclusions: enthusiastic and determined to know
although many of them were, they nevertheless

failed because they refused to tread the only path
which leads to true advances in natural science—
the path of unprejudiced accurate experiment, and
of careful reasoning on experimentally determined
facts.

And even when they had become convinced that
their aims were visionary, they could not break
free from the vicious system which bound them.

> ". . . I am broken and trained
> To my old habits: they are part of me.
> I know, and none so well, my darling ends
> Are proved impossible: no less, no less,
> Even now what humours me, fond fool, as when
> Their faint ghosts sit with me and flatter me,
> And send me back content to my dull round." *

One of the most commonly occurring and most
noticeable changes in the properties of matter is
that which proceeds when a piece of wood, or a
candle, or a quantity of oil burns. The solid wood,
or candle, or the liquid oil slowly disappears, and
this disappearance is attended with the visible formation of flame. Even the heavy fixed metals, tin
or lead, may be caused to burn; light is produced,
a part of the metal seems to disappear, and a white
(or reddish) solid, very different from the original
metal, remains. The process of burning presents
all those peculiarities which are fitted to strike an
observer of the changes of Nature; that is, which
are fitted to strike a chemist—for chemistry has

* Browning's " Paracelsus."

always been recognized as having for its object to explain the changes which matter undergoes. The chemists of the seventeenth and eighteenth centuries were chiefly occupied in trying to explain this process of burning or combustion.

Van Helmont (1577–1644), who was a physician and chemist of Brussels, clearly distinguished between common air and other "airs" or gases produced in different ways. Robert Hooke (1635–1703), one of the original Fellows of the Royal Society, in the "Micographia, or Philosophical Description of Minute Bodies," published in 1665, concluded from the results of numerous experiments that there exists in common air a peculiar kind of gas, similar to, or perhaps identical with the gas or air which is got by heating saltpetre ; and he further supposed that when a solid burns, it is dissolved by (or we should now say, it is converted into a gas by combining with) this peculiar constituént of the air.

John Mayow (1645–1679), a physician of Oxford, experimented on the basis of facts established by Hooke. He showed that when a substance, *e.g.* a candle, burns in air, the volume of air is thereby lessened. To that portion of the air which had *dissolved* the burned substance he gave the name of *nitre-air*, and he argued that this air exists in condensed form in nitre, because sulphur burns when heated with nitre in absence of common air. Mayow added the most important fact—a fact which was forgotten by many later experimenters— that the solid substance obtained by burning a metal in air weighs more than the metal itself did before burning. He explained this increase in weight by saying that the burning metal absorbs particles of " nitre-air " from the atmosphere. Thus Hooke and Mayow had really established the fact that common air consists of more than one definite kind of matter—in other words, that common air is not an element ; but until recent times the term " element " or " elementary principle " was used without any definite meaning. When we say that the ancients and the alchemists recognized four elements— earth, air, fire, and water—we do not attach to the word " element " the same definite meaning as when we now say, " Iron is an element."

From earth, air, fire and water other substances were obtained ; or it might be possible to resolve other substances into one or more of these four. But even to such a word as " substance " or " matter " no very definite meaning could be attached. Although, therefore, the facts set forth by Hooke and Mayow might now justify the assertion that air is not an element, they did not, in the year 1670, necessarily convey this meaning to men's minds. The distinction between element and compound was much more clearly laid down by the Hon. Robert Boyle (1627– 1691), whose chemical work was wonderfully accu- rate and thorough, and whose writings are charac- terized by acute scientific reasoning. We shall again return to these terms " element " and " compound."

But the visible and striking phenomenon in most processes of burning is the production of light and sometimes of flame. The importance of the fact that the burned substance (when a solid) weighs more than the unburned substance was over-shadowed by the apparent importance of the out-ward part of the process, which could scarcely be passed over by any observer. There appears to be an outrush of *something* from the burning substance. There *is* an outrush of something, said Becher and Stahl, and this something is the "principle of fire." The principle of fire, they said, is of a very subtle nature ; its particles, which are always in very rapid motion, can penetrate any substance, however dense. When metals burn—the argu-ment continued—they lose this principle of fire ; when the burned metal—or *calx* as it was usually called—is heated with charcoal it regains this "principle," and so the metal is re-formed from the calx.

Thus arose the famous theory of *phlogiston* (from Greek, = "burned"), which served as a central nucleus round which all chemical facts were grouped for nearly a hundred years.

John Joachim Becher was born at Speyer in 1635, and died in 1682 ; in his chemical works, the most important of which is the "Physica Subter-ranea," he retained the alchemical notion that the metals are composed of three "principles"—the nitrifiable, the combustible, and the mercurial—and taught that during calcination the combustible and mercurial principles are expelled, while the nitrifiable remains in the calx.

George Ernest Stahl—born at Anspach in 1660, and died at Berlin in 1734—had regard chiefly to the principles which escape during the calcination of metals, and simplifying, and at the same rendering more definite the idea of Becher, he conceived and enunciated the theory of phlogiston.

But if *something* (name it "phlogiston" or call it by any other name you please) is lost by a metal when the metal is burned, how is it that the loss of this thing is attended with an increase in the weight of the matter which loses it? Either the theory of phlogiston must be abandoned, or the properties of the *thing* called phlogiston must be very different from those of any known kind of matter.

Stahl replied, phlogiston is a "principle of levity;" the presence of phlogiston in a substance causes that substance to weigh less than it did before it received this phlogiston.

In criticizing this strange statement, we must remember that in the middle of the seventeenth century philosophers in general were not firmly convinced of the truth that the essential character of matter is that it possesses weight, nor of the truth that it is impossible to destroy or to create any quantity of matter however small. It was not until the experimental work of Lavoisier became generally known that chemists were convinced of these truths. Nevertheless, the opponents of the Stahlian doctrine were justified in asking for further explanations—in demanding that some other facts analogous to this supposed fact, viz. that a substance can weigh less than nothing, should be experimentally established.

The phlogistic theory however maintained its ground ; we shall find that it had a distinct element of truth in it, but we shall also find that it did harm to scientific advance. This theory was a wide and sweeping generalization from a few facts ; it certainly gave a central idea around which some facts might be grouped, and it was not very difficult, by slightly cutting down here and slightly adding there, to bring many new discoveries within the general theory.

We now know that in order to explain the process of combustion much more accurate knowledge was required than the chemists of the seventeenth century possessed ; but we ought to be thankful to these chemists, and notably to Stahl, that they did not hesitate to found a generalization on the knowledge they had. Almost everything propounded in natural science has been modified as man's knowledge of nature has become wider and more accurate ; but it is because the scientific student of nature uses the generalizations of to-day as stepping-stones to the better theories of to-morrow, that science grows "from more to more."

Looking at the state of chemistry about the middle of the eighteenth century, we find that the experiments, and especially the measurements, of Hooke and Mayow had laid a firm basis of fact concerning the process of combustion, but that the phlogistic theory, which appeared to contradict these facts, was supreme ; that the existence of airs, or gases, different from common air was established, but that the properties of these airs were very slightly and very inaccurately known ; that Boyle had distinguished element from compound and had given definite meanings to these terms, but that nevertheless the older and vaguer expression, "elementary principle," was generally used ; and lastly, that very few measurements of the masses of the different kinds of matter taking part in chemical changes had yet been made.

CHAPTER II.

ESTABLISHMENT OF CHEMISTRY AS A SCIENCE —PERIOD OF BLACK, PRIESTLEY AND LAVOISIER.

Joseph Black, 1728-1799. Joseph Priestley, 1733-1804. Antoine Laurent Lavoisier, 1743-1794.

DURING this period of advance, which may be broadly stated as comprising the last half of the eighteenth century, the aim and scope of chemical science were clearly indicated by the labours of Black, Priestley and Lavoisier. The work of these men dealt chiefly with the process of combustion. Black and Priestley finally proved the existence of airs or gases different from common air, and Lavoisier applied these discoveries to give a clear explanation of what happens when a substance burns.

JOSEPH BLACK was born near Bordeaux in the year 1728. His father was of Scottish family, but a native of Belfast; his mother was the daughter of Mr. Gordon, of Hilhead in Aberdeenshire. We are told by Dr. Robison, in his preface to Black's Lectures, that John Black, the father of Joseph, was a man "of most amiable manners, candid and liberal in his sentiments, and of no common information."

At the age of twelve Black was sent home to a school at Belfast; after spending six years there he went to the University of Glasgow in the year 1746. Little is known of his progress at school or at the university, but judging from his father's letters, which his son preserved, he seems to have devoted himself to study. While at Glasgow he was attracted to the pursuit of physical science, and chose medicine as a profession. Becoming a pupil of Dr. Cullen, he was much impressed with the importance of chemical knowledge to the student of medicine. Dr. Cullen appears to have been one of the first to take large and philosophical views of the scope of chemical science, and to attempt to raise chemistry from the rank of a useful art to that of a branch of natural philosophy. Such a man must have been attracted by the young student, whose work was already at once accurate in detail and wide in general scope.

In the notes of work kept by Black at this time are displayed those qualities of methodical arrangement, perseverance and thoroughness which are so prominent in his published investigations and lectures. In one place we find, says his biographer, many disjointed facts and records of diverse observations, but the next time he refers

to the same subjects we generally have analogous
facts noted and some conclusions drawn—we have
the beginnings of knowledge. Having once entered
on an investigation Black works it out steadily
until he gets definite results.

His earlier notes are concerned chiefly with heat
and cold ; about 1752 he begins to make references
to the subject of "fixed air."

About 1750 Black went to Edinburgh University
to complete his medical studies, and here he was
again fortunate in finding a really scientific student
occupying the chair of natural philosophy.

The attention of medical men was directed at
this time to the action of lime-water as a remedy
for stone in the bladder. All the medicines which
were of any avail in mitigating the pain attendant
on this disease more or less resembled the "caustic
ley of the soap-boilers" (or as we should now call
it caustic potash or soda). These caustic medicines
were mostly prepared by the action of quicklime on
some other substance, and quicklime was generally
supposed to derive its caustic, or corrosive properties from the fire which was used in changing
ordinary limestone into quicklime.

When quicklime was heated with "fixed alkalis"
(*i.e.* with potassium or sodium carbonate), it
changed these substances into caustic bodies which
had a corrosive action on animal matter ; hence it
was concluded that the quicklime had derived a
"power"—or some said had derived "igneous
matter"—from the fire, and had communicated this

to the fixed alkalis, which thereby acquired the property of corroding animal matter.

Black thought that he might be able to lay hold of this "igneous matter" supposed to be taken by the limestone from the fire ; but he found that limestone loses weight when changed into quicklime. He then dissolved limestone (or chalk) in spirits of salt (hydrochloric acid), and compared the loss of weight undergone by the chalk in this process with the loss suffered by an equal quantity of chalk when strongly heated. This investigation led Black to a fuller study of the action of heat on chalk and on "mild magnesia" (or as we now say, magnesium carbonate).

In order that his experiments might be complete and his conclusions well established, he delayed taking the degree of Doctor of Medicine for three years. He graduated as M.D. in 1755, and presented his thesis on " Magnesia Alba, Quicklime and other Alkaline Substances," which contained the results of what is probably the first accurately quantitative examination of a chemical action which we possess.

Black prepared mild magnesia (magnesium carbonate) by boiling together solutions of Epsom salts (magnesium sulphate) and fixed alkali (potassium carbonate). He showed that when mild magnesia is heated—

1. It is much decreased in bulk.

2. It loses weight (twelve parts become five, according to Black).

3. It does not precipitate lime from solutions

of that substance in acids (Black had already shown that mild magnesia does precipitate lime).

He then strongly heated a weighed quantity of mild magnesia in a retort connected with a receiver; a few drops of water were obtained in the receiver, but the magnesia lost six or seven times as much weight as the weight of the water produced. Black then recalls the experiments of Hales, wherein airs other than common air had been prepared, and concludes that the loss of weight noticed when mild magnesia is calcined is probably due to expulsion, by the heat, of some kind of air. Dissolving some of his mild magnesia in acid he noticed that effervescence occurred, and from this he concluded that the same air which, according to his hypothesis, is expelled by heat, is also driven out from the mild magnesia by the action of acid. He then proceeded to test this hypothesis. One hundred and twenty grains of mild magnesia were strongly calcined; the calcined matter, amounting to seventy grains, was dissolved in dilute oil of vitriol, and this solution was mixed with common fixed alkali (potassium carbonate). The solid which was thus produced was collected, washed and weighed; it amounted to a trifle less than one hundred and twenty grains, and possessed all the properties—detailed by Black—of the original mild magnesia. But this is exactly the result which ought to have occurred according to his hypothesis.

The next step in the investigation was to collect the peculiar air which Black had proved to be

evolved during the calcination of mild magnesia.
To this substance he gave the name of " fixed air,"
because it was fixed or held by magnesia.   Black
established the existence of this air in the expired
breath of animals, and also showed that it was
present in the air evolved during vinous fermentation.   He demonstrated several of its properties ;
among these, the fact that animals die when placed
in this air.   An air with similar properties was
obtained by calcining chalk.   Black held that
the chemical changes which occur when chalk
is calcined are exactly analogous to those which
he had proved to take place when magnesia
is strongly heated.   Chalk ought therefore to
lose weight when calcined ; the residue ought to
neutralize an acid without evolution of any gas,
and the quantity of acid thus neutralized ought to
be the same as would be neutralized by the uncalcined chalk ; lastly, it ought to be possible to
recover the uncalcined chalk by adding a fixed
alkali to a solution of the calcined chalk or
quicklime.

The actual results which Black obtained were as
follows :—

One hundred and twenty grains of chalk were
dissolved in dilute muriatic (hydrochloric) acid ;
421 grains of the acid were needed to neutralize
the chalk, and 48 grains of fixed air were evolved.
One hundred and twenty grains of the same
specimen of chalk were strongly calcined, and
then dissolved in dilute muriatic acid ; 414 grains

of the acid were required to neutralize the calcined chalk. The difference between 421 and 414 is very slight ; considering the state of practical chemistry at Black's time, we may well agree with him that he was justified in the conclusion that equal weights of calcined and of uncalcined chalk neutralize the same amount of acid. One hundred and twenty grains of the same specimen of chalk were again strongly heated ; the calcined chalk, amounting to 68 grains, was digested with a solution of fixed alkali in water. The substance thus obtained, when washed and dried, weighed 118 grains, and had all the properties of ordinary chalk. Therefore, said Black, it is possible to recover the whole of the chalk originally present before calcination, by adding a fixed alkali to the calcined chalk or quicklime.

At this time it was known that water dissolves quicklime, but it was generally held that only about one-fourth (or perhaps a little more) of any specimen of quicklime could be dissolved by water, however much water was employed. Black's researches had led him to regard quicklime as a homogeneous chemical compound ; he concluded that as water undoubtedly dissolves quicklime to some extent, any specimen of this substance, provided it be pure, must be wholly soluble in water. Carefully conducted experiments proved that Black's conclusion was correct. Black had thus proved that quicklime is a definite substance, with certain fixed properties which characterize it and mark it off from all other substances; that by absorbing, or combining with another definite substance (fixed air), quicklime is changed into a third substance, namely chalk, which is also characterized by properties as definite and marked as those of quicklime or fixed air.

Black, quite as much as the alchemists, recognized the fact that change is continually proceeding in Nature; but he clearly established the all-important conclusion that these natural changes proceed in definite order, and that it is possible by careful experiment and just reasoning to acquire a knowledge of this order.   He began the great work of showing that, as in other branches of natural science, so also in chemistry, which is pre-eminently the study of the changes of Nature, "the only distinct meaning of that word" (natural) "is *stated, fixed,* or *settled*" (Butler's "Analogy," published 1736).

This research by Black is a model of what scientific work ought to be.   He begins with a few observations of some natural phenomenon; these he supplements by careful experiments, and thus establishes a sure basis of fact; he then builds on this basis a general hypothesis, which he proceeds to test by deducing from it certain necessary conclusions, and proving, or disproving, these by an appeal to Nature.   This is the scientific method; it is common sense made accurate.

Very shortly after the publication of the thesis on magnesia and quicklime, a vacancy occurred in the chemical chair in Glasgow University, and Black was appointed Professor of Anatomy and Lecturer on Chemistry. As he did not feel fully qualified to lecture on anatomy, he made an arrangement to exchange subjects with the Professor of Medicine, and from this time he delivered lectures on chemistry and on "The Institutes of Medicine."

Black devoted a great deal of care and time to the teaching duties of his chair. His chemical experimental researches were not much advanced after this time ; but he delivered courses of lectures in which new light was thrown on the whole range of chemical science.

In the years between 1759 and 1763 Black examined the phenomena of heat and cold, and gave an explanation, founded on accurate experiments, of the thermal changes which accompany the melting of solids and the vaporization of liquids.

If pieces of wood, lead and ice be taken by the hand from a box in which they have been kept cold, the wood feels cold to the touch, the lead feels colder than the wood, and the ice feels colder than the lead ; hence it was concluded that the hand receives cold from the wood, more cold from the lead, and most cold from the ice.

Black however showed that the wood really takes away heat from the hand, but that as the wood soon gets warmed, the process stops before long ; that the lead, not being so quickly warmed as the wood, takes away more heat from the hand than the wood does, and that the ice takes away more heat than either wood or lead.

Black thought that the heat which is taken by melting ice from a warm body remains in the water which is produced ; as soon as winter came he proceeded to test this supposition by comparing the times required to melt one pound of ice and to raise the temperature of one pound of water through one degree, the source of heat being the same in each case. He also compared the time required to lower the temperature of one pound of water through one degree with that required to freeze one pound of ice-cold water. He found that in order to melt one pound of ice without raising its temperature, as much heat had to be added to the ice as sufficed to raise the temperature of one pound of water through about 140 degrees of Fahrenheit's thermometer. But this heat which has been added to the ice to convert it into water is not indicated by the thermometer. Black called this " *latent heat.*"

The experimental data and the complete theory of latent heat were contained in a paper read by Black to a private society which met in the University of Glasgow, on April 23, 1762 ; but it appears that Black was accustomed to teach the theory in his ordinary lectures before this date.

The theory of latent heat ought also to explain the phenomena noticed when liquid water is changed into steam. Black applied his theory generally to this change, but did not fully work out the details and actually measure the quantity of heat which is absorbed by water at the boiling point before it is wholly converted into steam at the same temperature, until some years later when he had the assistance of his pupil and friend James Watt.

Taking a survey of the phenomena of Nature, Black insisted on the importance of these experimentally established facts—that before ice melts it must absorb a large quantity of heat, and before water is vaporized it must absorb another large quantity of heat, which amounts of heat are restored to surrounding substances when water vapour again becomes liquid water and when liquid water is congealed to ice. He allows his imagination to picture the effects of these properties of water in modifying and ameliorating the climates of tropical and of Northern countries. In his lectures he says, " Here we can also trace another magnificent train of changes which are nicely accommodated to the wants of the inhabitants of this globe. In the equatorial regions, the oppressive heat of the sun is prevented from a destructive accumulation by copious evaporation. The waters, stored with their vaporific heat, are then carried aloft into the atmosphere till the rarest of the vapour reaches the very cold regions of the air, which immediately forms a small portion of it into a fleecy cloud. This also further tempers the scorching heat by its opacity, performing the acceptable office of a screen. From thence the clouds are carried to the inland countries, to form the sources in the mountains which are to supply the numberless streams that water the fields. And by the steady operation of causes, which are tolerably uniform, the greater part of the vapours passes on to the circumpolar regions, there to descend in rains and dews; and by this beneficent conversion into rain by the cold of those regions, each particle of steam gives up the heat which was latent in it. This is immediately diffused, and softens the rigour of those less comfortable climates."

In the year 1766 Black was appointed Professor of Chemistry in the University of Edinburgh, in which position he remained till his death in 1799. During these thirty-three years he devoted himself chiefly to teaching and to encouraging the advance of chemical science. He was especially careful in the preparation of his elementary lectures, being persuaded that it was of the utmost importance that his pupils should be well grounded in the principles of chemistry.

His health had never been robust, and as he grew old he was obliged to use great care in his diet; his simple and methodical character and habits made it easy for him to live on the plainest food, and to take meals and exercise at stated times and in fixed quantities.

Black's life closed, as was fitting, in a quiet and honoured old age. He had many friends, but lived pretty much alone—he was never married.

On the 26th of November 1799, " being at table with his usual fare, some bread, a few prunes and a measured quantity of milk diluted with water, and having the cup in his hand when the last stroke of his pulse was to be given, he had set it down on his knees, which were joined together, and kept it steady with his hand, in the manner of a person perfectly at ease; and in this attitude he expired, without spilling a drop, and without a writhe in his countenance, as if an experiment had been required to show to his friends the facility with which he departed."

Black was characterized by "moderation and sobriety of thought;" he had a great sense of the fitness of things—of what is called by the older writers "propriety." But he was by no means a dull companion; he enjoyed general society, and was able to bear a part in any kind of conversation. A thorough student of Nature, he none the less did not wish to devote his whole time to laboratory work or to the labours of study; indeed he seems to have preferred the society of well-cultivated men and women to that of specialists in his own or other branches of natural science. But with his true scientific peers he doubtless appeared at his best. Among his more intimate friends were the famous political economist Adam Smith, and the no less celebrated philosopher David Hume. Dr. Hutton, one of the earliest workers in geology, was a particular friend of Black; his friendship with James Watt began when Watt was a student in his class, and continued during his life.

With such men as his friends, and engaged in the study of Nature—that boundless subject which one can never know to the full, but which one can always know a little more year by year—Black's life could not but be happy. His example and his teaching animated his students; he was what a university professor ought to be, a student among students, but yet a teacher among pupils. His work gained for him a place in the first rank of men of science; his clearness of mind, his moderation, his gentleness, his readiness to accept the views of others provided these views were well established on a basis of experimentally determined facts, fitted him to be the centre of a circle of scientific students who looked on him as at once their teacher and their friend.

As a lecturer Black was eminently successful. He endeavoured to make all his lectures plain and intelligible; he enlivened them by many experiments designed simply to illustrate the special point which he had in view. He abhorred ostentatious display and trickiness in a teacher.

Black was strongly opposed to the use of hypotheses in science. Dr. Robison (the editor of his lectures) tells that when a student in Edinburgh he met Black, who became interested in him from hearing him speak somewhat enthusiastically in favour of one of the lecturers in the university. Black impressed on him the necessity of steady experimental work in natural science, gave him a copy of Newton's "Optics" as a model after which scientific work ought to be conducted, and advised him "to reject, even without examination, any hypothetical explanation, as a mere waste of time and ingenuity." But, when we examine Black's own work, we see that by "hypothetical explanations" he meant vague guesses. He himself made free use of scientific (*i.e.* of exact) hypotheses ; indeed the history of science tells us that without hypotheses advance is impossible. Black taught by his own researches that science is not an array of facts, but that the object of the student of Nature is to explain facts. But the method generally in vogue before the time of Black was to gather together a few facts, or what seemed to be facts, and on these to raise a vast superstructure of "vain imaginings." Naturalists had scarcely yet learned that Nature is very complex, and that guessing and reasoning on guesses, with here and there an observation added, was not the method by which progress was to be made in learning the lessons written in this complex book of Nature.

In place of this loose and slipshod method Black insisted that the student must endeavour to form a clear mental image of every phenomenon which he studied. Such an image could be obtained only by beginning with detailed observation and experiment. From a number of definite mental images the student must put together a picture of the whole natural phenomenon under examination ; perceiving that something was wanted here, or that the picture was overcrowded there, he must again go to Nature and gain fresh facts, or sometimes prove that what had been accepted as facts had no real existence, and so at length he would arrive at a true representation of the whole process.

So anxious was Black to define clearly what he knew and professed to teach, that he preferred to call his lectures "On the Effects of Heat and Mixtures," rather than to announce them as "A Systematic Course on Chemistry."

His introductory lecture on "Heat in General" is very admirable ; the following quotation will serve to show the clearness of his style and the methodical but yet eminently suggestive manner of his teaching :—

"Of Heat in General.

"That this extensive subject may be treated in a profitable manner, I propose—

"First. To ascertain what I mean by the word *heat* in these lectures.

"Secondly. To explain the meaning of the term *cold*, and ascertain the real difference between heat and cold.

"Thirdly. To mention some of the attempts which have been made to discover the nature of heat, or to form an idea of what may be the immediate cause of it.

"Fourthly and lastly. I shall begin to describe sensible effects produced by heat on the bodies to which it is communicated.

"Any person who reflects on the ideas which we annex to the word *heat* will perceive that this word is used for two meanings, or to express two different things. It either means a sensation excited in our organs, or a certain quality, affection, or condition of the bodies around us, by which they excite in us that sensation. The word is used in the first sense when we say, we feel heat; in the second, when we say, there is heat in the fire or in a hot stone. There cannot be a sensation of heat in the fire, or in the hot stone, but the matter of the fire, or of the stone, is in a state or condition by which it excites in us the sensation of heat.

" Now, in beginning to treat of heat and its effects, I propose to use the word in this second sense only; or as expressing that state, condition, or quality of matter by which it excites in us the sensation of heat. This idea of heat will be modified a little and extended as we proceed, but the meaning of the word will continue at bottom the same, and the reason of the modification will be easily perceived."

Black's manner of dealing with the phenomenon of combustion illustrates the clearness of the conceptions which he formed of natural phenomena, and shows moreover the thoroughly unbiased nature of his mind. As soon as he had convinced himself that the balance of evidence was in favour of the new (antiphlogistic) theory, he gave up those doctrines in which he had been trained, and accepted the teaching of the French chemists; but he did not—as some with less well-balanced minds might do—regard the new theory as a final statement, but rather as one stage nearer the complete explanation which future experiments and future reasoning would serve to establish.

In his lectures on combustion Black first of all establishes the facts, that when a body is burned it is changed into a kind (or kinds) of matter which is no longer inflammable ; that the presence of air is needed for combustion to proceed ; that the substance must be heated "to a certain degree" before combustion or inflammation begins; that this degree of heat (or we should now say this degree of temperature) differs for each combustible substance ; that the supply of air must be renewed if the burning is to continue ; and that the process of burning produces a change in the quality of the air supplied to the burning body.

He then states the phlogistic interpretation of these phenomena : that combustion is caused by the outrush from the burning body of a something called the *principle of fire*, or *phlogiston*.

Black then proceeds to demonstrate certain other facts :—When the substances produced by burning phosphorus or sulphur are heated with carbon (charcoal) the original phosphorus or sulphur is reproduced. This reproduction is due, according to the phlogistic chemists, to the giving back, by carbon, of the phlogiston which had escaped during the burning. Hence carbon contains much phlogiston. But as a similar reproduction of phosphorus or sulphur, from the substances obtained by burning these bodies, can be accomplished by the use of substances other than carbon, it is evident that these other substances also contain much phlogiston, and, moreover, that the phlogiston contained in all these substances is one and the same *principle*. What then, he asks, is this "principle" which can so escape, and be so restored by the action of various substances? He then proceeds as follows :—

"But when we inquire further, and endeavour to learn what notion was formed of the nature of this principle, and what qualities it was supposed to have in its separate state, we find this part of the subject very obscure and unsatisfactory, and the opinions very unsettled.

"The elder chemists, and the alchemists, considered sulphur as the universal inflammable principle, or at least they chose to call the inflammable part of all bodies, that are more or less inflammable, by the name of their sulphur. . . . The famous German chemist Becher was, I believe, the first who rejected the notion of sulphur being the principle of inflammability in bodies. . . . His notion of the nature of the pure principle of inflammability was afterwards more fully explained and supported by Professor Stahl, who, agreeably to the doctrine of Becher, represented the principle of inflammability as a dry substance, or of an earthy nature, the particles of which were exquisitely subtile, and were much disposed to be agitated and set in motion with inconceivable velocity. . . . The opinion of Becher and Stahl concerning this *terra secunda*, or *terra inflammabilis*, or *phlogis-*

ton, was that the atoms of it are, more than all others, disposed to be affected with an excessively swift whirling motion (*motus vorticillaris*). The particles of other elementary substances are likewise liable to be affected with the same sort of motion, but not so liable as those of *terra secunda ;* and when the particles of any body are agitated with this sort of motion, the body exhibits the phenomena of heat, or ignition, or inflammation according to the violence and rapidity of the motion. . . . Becher and Stahl, therefore, did not suppose that heat depended on the abundance of a peculiar matter, such as the matter of heat or fire is now supposed to be, but on a peculiar motion of the particles of matter. . . .

"This very crude opinion of the earthy nature of the principle of inflammability appears to have been deduced from a quality of many of the inflammable substances, by which they resist the action of water as a solvent. The greater number of the earthy substances are little, or not at all, soluble in water. . . . And when Becher and Stahl found those compounds, which they supposed contained phlogiston in the largest quantity, to be insoluble in water, although the other matter, with which the phlogiston was supposed to be united, was, in its separate state, exceedingly soluble in that fluid, they concluded that *a dry nature*, or *an incapability to be combined with water*, was an eminent quality of their phlogiston ; and this was what they meant by calling it an earth or earthy substance.

III.   E

. . . But these authors supposed, at the same time, that the particles of this dry and earthy phlogiston were much disposed to be excessively agitated with a whirling motion ; which whirling motion, exerted in all directions from the bodies in which phlogiston is contained, produced the phenomena of inflammation. This appears to have been the notion formed by Becher and Stahl, concerning the nature of the principle of inflammability, or the phlogiston ; a notion which seems the least entitled to the name of explanation of anything we can think of. I presume that few persons can form any clear conception of this whirling motion, or, if they can, are able to explain to themselves how it produces, or can produce, anything like the phenomena of heat or fire."

Black then gives a clear account of the experiments of Priestley and Lavoisier (see pp. 58, 59, and 87–89), which established the presence, in common air, of a peculiar kind of gas which is especially concerned in the processes of combustion ; he emphasizes the fact that a substance increases in weight when it is burned ; and he gives a simple and clear statement of that explanation of combustion which is now accepted by all, and which does not require that the existence of any principle of fire should be assumed.

It is important to note that Black clearly connects the *physical* fact that heat is absorbed, or evolved, by a substance during combustion, with the *chemical* changes which are brought about in the properties of the substance burned. He concludes with an admirable contrast between the phlogistic theory and the theory of Lavoisier, which shows how wide, and at the same time how definite, his conceptions were. Black never speaks contemptuously of a theory which he opposes.

"According to this theory" (*i.e.* the theory of Lavoisier), "the inflammable bodies, sulphur for example, or phosphorus, are simple substances. The acid into which they are changed by inflammation is a compound. The chemists, on the contrary" (*i.e.* the followers of Stahl), "consider the inflammable bodies as compounds, and the uninflammable matter as more simple. In the common theory the heat and light are supposed to emanate from, or to be furnished by, the burning body. But, in Mr. Lavoisier's theory, both are held to be furnished by the air, of which they are held to be constituent parts, or ingredients, while in its state of fire-supporting air."

Black was not a brilliant discoverer, but an eminently sound and at the same time imaginative worker; whatever he did he did well, but he did not exhaust any field of inquiry. Many of the facts established by him have served as the basis of important work done by those who came after him. The number of new facts added by Black to the data of chemistry was not large; but by his lectures—which are original dissertations of the highest value—he did splendid service in advancing the science of chemistry. Black possessed that which has generally distinguished great men of science, a marked honesty of character ; and to this he added comprehensiveness of mental vision : he saw beyond the limits of the facts which formed the foundations of chemical science in his day. He was not a fact-collector, but a philosopher.

JOSEPH PRIESTLEY, the son of Jonas Priestley, "a maker and dresser of woollen cloth," was born at Fieldhead, near Leeds, in the year 1733. His mother, who was the daughter of a farmer near Wakefield, died when he was seven years old. From that time he was brought up by a sister of his father, who was possessed of considerable private means.

Priestley's surroundings in his young days were decidedly religious, and evidently gave a tone to his whole after life. We shall find that Priestley's work as a man of science can scarcely be separated from his theological and metaphysical work. His cast of mind was decidedly metaphysical ; he was altogether different from Black, who, as we have seen, was a typical student of natural phenomena.

The house of Priestley's aunt was a resort for all the Dissenting ministers of that part of the county. She herself was strictly Calvinistic in her theological views, but not wholly illiberal.

Priestley's early schooling was chiefly devoted to learning languages ; he acquired a fair knowledge of Latin, a little Greek, and somewhat later he learned the elements of Hebrew. At one time he thought of going into trade, and therefore, as he

tells us in his " Memoirs," he acquired some knowledge of French, Italian and High Dutch. With
the help of a friend, a Dissenting minister, he
learned something of geometry, mathematics and
natural philosophy, and also got some smattering
of the Chaldee and Syriac tongues.

At the age of nineteen Priestley went to an
"academy" at Daventry. The intellectual atmosphere here seems to have been suitable to the
rapid development of Priestley's mind. Great
freedom of discussion was allowed; even during
the teachers' lectures the students were permitted
"to ask whatever questions and to make whatever remarks" they pleased; and they did it,
Priestley says, "with the greatest, but without any
offensive, freedom."

The students were required to read and to give
an account of the more important arguments for
and against the questions discussed in the teachers'
lectures. Theological disputations appear to have
been the favourite topics on which the students
exercised their ingenuity among themselves.
Priestley tells us that he "saw reason to embrace
what is generally called the heterodox side of
almost every question."

Leaving this academy, Priestley went, in 1755,
as assistant to the Dissenting minister at Needham,
in Suffolk. Here he remained for three years, living
on a salary of about £30 a year, and getting more
and more into bad odour because of his peculiar
theological views.

From Needham he moved to Nantwich, in Cheshire, where he was more comfortable, and, having plenty of work to do, he had little time for abstruse speculations. School work engaged most of his time at Nantwich; he also began to collect a few scientific instruments, such as an electrical machine and an air-pump. These he taught his scholars to use and to keep in good order. He gave lectures on natural phenomena, and encouraged his scholars to make experiments and sometimes to exhibit their experiments before their parents and friends. He thus extended the reputation of his school and implanted in his scholars a love of natural knowledge.

In the year 1761 Priestley removed to Warrington, to act as tutor in a newly established academy, where he taught languages—a somewhat wide subject, as it included lectures on " The Theory of Languages," on " Oratory and Criticism," and on " The History, Laws, and Constitution of England." He says, " It was my province to teach elocution, and also logic and Hebrew. The first of these I retained, but after a year or two I exchanged the two last articles with Dr. Aikin for the civil law, and one year I gave a course of lectures on anatomy."

During his stay at Warrington, which lasted until 1767, Priestley married a daughter of Mr. Isaac Wilkinson, an ironmaster of Wrexham, in Wales. He describes his wife as " a woman of an excellent understanding much improved by reading, of great fortitude and strength of mind, and of a temper in the highest degree affectionate and generous, feeling strongly for others and little for herself, also greatly excelling in everything relating to household affairs."

About this time Priestley met Dr. Franklin more than once in London. His conversation seems to have incited Priestley to a further study of natural philosophy. He began to examine electrical phenomena, and this led to his writing and publishing a " History of Electricity," in the course of which he found it necessary to make new experiments. The publication of the results of these experiments brought him more into notice among scientific men, and led to his election as a Fellow of the Royal Society, and to his obtaining the degree of LL.D. from the University of Edinburgh. In the year 1767 Priestley removed to Leeds, where he spent six years as minister of Millhill Chapel.

He was able to give freer expression to his theological views in Leeds than could be done in smaller places, such as Needham and Nantwich. During this time he wrote and published many theological and metaphysical treatises. But, what is of more importance to us, he happened to live near a brewery. Now, the accidental circumstances, as we call them, of Priestley's life were frequently of the greatest importance in their effects on his scientific work. Black had established the existence and leading properties of fixed air about twelve or thirteen years before the time when Priestley came

to live near the brewery in Leeds. He had shown
that this fixed air is produced during alcoholic fermentation. Priestley knowing this used to collect
the fixed air which came off from the vats in the
neighbouring brewery, and amuse himself with observing its properties. But removing from this part
of the town his supplies of fixed air were stopped.
As however he had become interested in working
with airs, he began to make fixed air for himself from
chalk, and in order to collect this air he devised
a very simple piece of apparatus which has played
a most important part in the later development of
the chemistry of gases, or pneumatic chemistry.
Priestley's *pneumatic trough* is at this day to be
found in every laboratory; it is extremely simple
and extremely perfect. A dish of glass, or earthen-
ware, or wood is partly filled with water; a shelf
runs across the dish at a little distance beneath
the surface of the water; a wide-mouthed bottle is
filled with water and placed, mouth downwards,
over a hole in this shelf. The gas which is to be
collected in this bottle is generated in a suitable
vessel, from which a piece of glass or metal tubing
passes under the shelf and stops just where the
hole is made. The gas which comes from the apparatus bubbles up into the bottle, drives out the
water, and fills the bottle. When the bottle is full
of gas, it is moved to one side along the shelf, and
another bottle filled with water is put in its place.
As the mouth of each bottle is under water there
is no connection between the gas inside and the

air outside the bottle ; the gas may therefore be kept in the bottle until the experimenter wants it. (See Fig. 1. which is reduced from the cut in Priestley's " Air.")

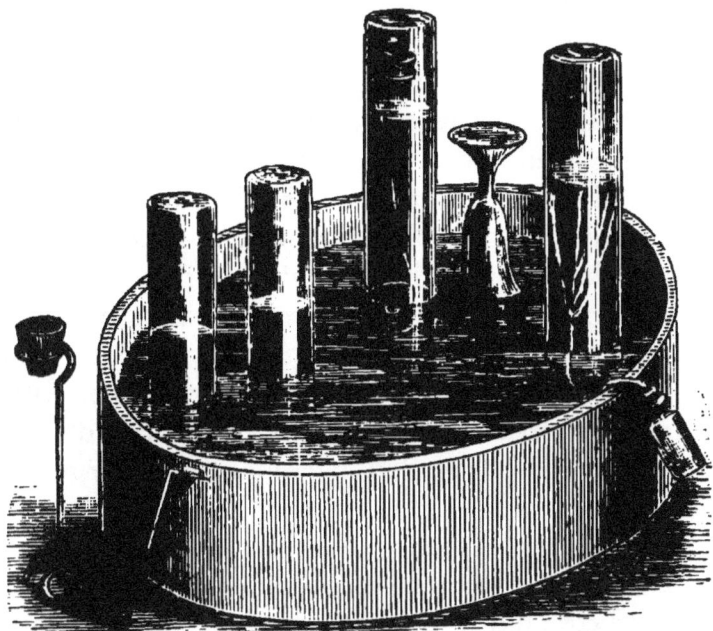

Fig. 1.

Priestley tells us that at this time he knew very little chemistry, but he thinks that this was a good thing, else he might not have been led to make so many new discoveries as he did afterwards make.

Experimenting with fixed air, he found that water could be caused to dissolve some of the gas. In 1772 he published a pamphlet on the method of

impregnating water with fixed air; this solution
of fixed air in water was employed medicinally,
and from this time we date the manufacture of
artificial mineral waters.

The next six years of Priestley's life (1773–1779)
are very important in the history of chemistry; it
was during these years that much of his best work
on various airs was performed.   During this
time he lived as a kind of literary companion
(nominally as librarian) with the Earl of Shelburne
(afterwards Marquis of Lansdowne.)   His wife and
family—he had now three children—lived at Calne,
in Wiltshire, near Lord Shelburne's seat of Bowood.
Priestley spent most of the summer months with
his family, and the greater part of each winter with
Lord Shelburne at his London residence; during
this time he also travelled in Holland and Germany,
and visited Paris in 1774.

In a paper published in November 1772, Priestley
says that he examined a specimen of air which
he had extracted from saltpetre above a year before
this date.   This air "had by some means or other
become noxious, but," he supposed, "had been restored to its former wholesome state, so as to
effervesce with nitrous air" (in modern language,
to combine with nitric oxide) "and to admit a
candle to burn in it, in consequence of agitation
with water."   He tells us, in his "Observations on
Air" (1779), that at this time he was altogether in
the dark as to the nature of this air obtained from
saltpetre.   In August 1774, he was amusing himself

by observing the action of heat on various substances—"without any particular view," he says,
"except that of extracting air from a variety of
substances by means of a burning lens in quicksilver,
which was then a new process with me, and which
I was very proud of"—when he obtained from *red
precipitate* (oxide of mercury) an air in which a
candle burned with a "remarkably vigorous flame."
The production of this peculiar air "surprised me
more than I can well express;" "I was utterly at a
loss how to account for it." At first he thought
that the specimen of *red precipitate* from which the
air had been obtained was not a proper preparation, but getting fresh specimens of this salt, he
found that they all yielded the same kind of air.
Having satisfied himself by experiment that this
peculiar air had "all the properties of common air,
only in much greater perfection," he gave to it the
name of *dephlogisticated air.* Later experiments
taught him that the same air might be obtained
from red lead, from manganese oxide, etc., by the
action of heat, and from various other salts by the
action of acids.

Priestley evidently regards the new " dephlogisticated air " simply as very pure ordinary air ; indeed,
he seems to look on all airs, or gases, as easily
changeable one into the other. He always inter-
prets his experimental results by the help of the
theory of phlogiston. One would indeed think from
Priestley's papers that the existence of this substance phlogiston was an unquestioned and unques-

tionable fact. Thus, he says in the preface to his
"Experiments on Air:" "If any opinion in all the
modern doctrine concerning air be well founded, it
is certainly this, that nitrous air is highly charged
with phlogiston, and that from this quality only it
renders pure air noxious. . . . If I have completely
ascertained anything at all relating to air it is
this." Priestley thought that "very pure air"
would take away phlogiston from some metals
without the help of heat or any acid, and thus cause
these metals to rust. He therefore placed some clean
iron nails in *dephlogisticated air* standing over mercury; after three months he noticed that about one-
tenth of the air in the vessel had disappeared, and
he concluded, although no rust appeared, that the
dephlogisticated air had as a fact withdrawn phlogiston from the iron nails. This is the kind of
reasoning which Black described to his pupils as
"mere waste of time and ingenuity." The experiment with the nails was made in 1779; at this time,
therefore, Priestley had no conception as to what
his *dephlogisticated air* really was.

Trying a great many experiments, and finding
that the new air was obtained by the action of acids
on earthy substances, Priestley was inclined to
regard this air, and if this then all other airs, as
made up of an acid (or acids) and an earthy substance. We now know how completely erroneous
this conclusion was, but we must remember that in
Priestley's time chemical substances were generally
regarded as of no very definite or fixed composition;

that almost any substance, it was supposed, might be changed into almost any other ; that no clear meaning was attached to the word "element;" and that few, if any, careful measurements of the quantities of different kinds of matter taking part in chemical actions had yet been made.

But at the same time we cannot forget that the books of Hooke and Mayow had been published years before this time, and that twenty years before Priestley began his work on airs, Black had published his exact, scientific investigation on fixed air.

Although we may agree with Priestley that, had he made himself acquainted with what others had done before he began his own experiments, he might not have made so many new discoveries as he did, yet one cannot but think that his discoveries, although fewer, would have been more accurate.

We are told by Priestley that, when he was in Paris in 1774, he exhibited the method of obtaining dephlogisticated air from *red precipitate* to Lavoisier and other French chemists.   We shall see hereafter what important results to science followed from this visit to Lavoisier.

Let us shortly review Priestley's answer to the question, " What happens when a substance burns in air ? "

Beginning to make chemical experiments when he had no knowledge of chemistry, and being an extremely rapid worker and thinker, he naturally adopted the prevalent theory, and as naturally interpreted the facts which he discovered in accordance with this theory.

When a substance burns, phlogiston, it was said, rushes out of it. But why does rapid burning only take place in air? Because, said Priestley, air has a great affinity for phlogiston, and draws it out of the burning substance. What then becomes of this phlogiston? we next inquire. The answer is, obviously it remains in the air around the burning body, and this is proved by the fact that this air soon becomes incapable of supporting the process of burning, it becomes phlogisticated. Now, if phlogisticated air cannot support combustion, the greater the quantity of phlogiston in air, the less will it support burning; but we know that if a substance is burnt in a closed tube containing air, the air which remains when the burning is quite finished at once extinguishes a lighted candle. Priestley also proved that an air can be obtained by heating *red precipitate*, characterized by its power of supporting combustion with great vigour. What is this but common air completely deprived of phlogiston? It is dephlogisticated air. Now, if common air draws phlogiston out of substances, surely this dephlogisticated air will even more readily do the same. That it really does this Priestley thought he had proved by his experiment with clean iron nails (see p. 60).

Water was regarded as a substance which, like air, readily combined with phlogiston; but Priestley thought that a candle burned less vigorously in dephlogisticated air which had been shaken with water than in the same air before this treatment; hence he concluded that phlogiston had been taken from the water.

After Cavendish had discovered (or rather re-discovered) hydrogen, and had established the fact that this air is extremely inflammable, most chemists began to regard this gas as pure or nearly pure phlogiston, or, at least, as a substance very highly charged with phlogiston. "Now," said Priestley, "when a metal burns phlogiston rushes out of it; if I restore this phlogiston to the metallic calx, I shall convert it back into the metal." He then showed by experiment that when calx of iron is heated with hydrogen, the hydrogen disappears and the metal iron is produced.

He seemed, therefore, to have a large experimental basis for his answer to the question, "What happens when a substance burns?" But at a later time it was proved that iron was also produced by heating the calx of iron with carbon. The antiphlogistic chemists regarded fixed air as composed of carbon and dephlogisticated air; the phlogisteans said it was a substance highly charged with phlogiston. The antiphlogistic school said that calx of iron is composed of iron and dephlogisticated air; the phlogisteans said it was iron deprived of its phlogiston. Here was surely an opportunity for a crucial experiment: when calx of iron is heated with carbon, and iron is produced, there must either be a production of fixed air (which is a non-inflammable gas, and forms a white solid substance

when brought into contact with limewater), or
there must be an outrush of phlogiston from the
carbon. The experiment was tried: a gas was
produced which had no action on limewater and
which was very inflammable; what could this be
but phlogiston, already recognized by this very
property of extreme inflammability? Thus the
phlogisteans appeared to triumph. But if we examine these experiments made by Priestley with the
light thrown on them by subsequent research, we
find that they bear the interpretation which he put
on them only because they were not accurate; thus,
two gases are inflammable, but it by no means
follows that these gases are one and the same. We
must have more accurate knowledge of the properties of these gases.

The air around a burning body, such as iron,
after a time loses the power of supporting combustion; but this is merely a qualitative fact.
Accurately to trace the change in the properties
of this air, it is absolutely necessary that exact
measurements should be made; when this is done,
we find that the volume of air diminishes during
the combustion, that the burning body gains weight,
and that this gain in weight is just equal to the
loss in weight undergone by the air. When the
inflammable gas produced by heating calx of iron
with carbon was carefully and *quantitatively*
analyzed, it was found to consist of carbon and
oxygen (dephlogisticated air), but to contain these
substances in a proportion different from that in

which they existed in fixed air. It was a new kind of air or gas ; it was *not* hydrogen.

This account of Priestley's experiments and conclusions regarding combustion shows how easy it is in natural science to interpret experimental results, especially when these results are not very accurate, in accordance with a favourite theory ; and it also illustrates one of the lessons so emphatically taught by all scientific study, viz. the necessity of suspending one's judgment until accurate measurements have been made, and the great wisdom of then judging cautiously.

About 1779 Priestley left Lord Shelburne, and went as minister of a chapel to Birmingham, where he remained until 1791.

During his stay in Birmingham, Priestley had a considerable amount of pecuniary help from his friends. He had from Lord Shelburne, according to an agreement made when he entered his service, an annuity of £150 a year for life ; some of his friends raised a sum of money annually for him, in order that he might be able to prosecute his researches without the necessity of taking pupils. During the ten years or so after he settled in Birmingham, Priestley did a great deal of chemical work, and made many discoveries, almost entirely in the field of pneumatic chemistry.

Besides the discovery of dephlogisticated air (or oxygen) which has been already described, Priestley discovered and gave some account of the properties of *nitrous air* (nitric acid), *vitriolic acid*

III.   F *air* (sulphur dioxide), *muriatic acid air* (hydrochloric acid), and *alkaline air* (ammonia), etc.

In the course of his researches on the last-named air he showed, that when a succession of electric sparks is passed through this gas a great increase in the volume of the gas occurs. This fact was further examined at a later time by Berthollet, who, by measuring the increase in volume undergone by a measured quantity of ammonia gas, and determining the nature of the gases produced by the passage of the electric sparks, proved that ammonia is a compound of hydrogen and nitrogen, and that three volumes of the former gas combine with one volume of the latter to produce two volumes of ammonia gas.

Priestley's experiments on "inflammable air"—or hydrogen—are important and interesting. The existence of this substance as a definite kind of air had been proved by the accurate researches of Cavendish in 1766. Priestley drew attention to many actions in which this inflammable air is produced, chiefly to those which take place between acids and metals. He showed that inflammable air is not decomposed by electric sparks; but he thought that it was decomposed by long-continued heating in closed tubes made of lead-glass. Priestley regarded inflammable air as an air containing much phlogiston. He found that tubes of lead-glass, filled with this air, were blackened when strongly heated for a long time, and he explained this by saying that the lead in the glass had a great affinity for phlogiston, and drew it out of the inflammable air.

When inflammable air burns in a closed vessel containing common air, the latter after a time loses its property of supporting combustion. Priestley gave what appeared to be a fairly good explanation of this fact, when he said that the inflammable air parted with phlogiston, which, becoming mixed with the ordinary air in the vessel, rendered it unable to support the burning of a candle. He gave a few measurements in support of this explanation ; but we now know that the method of analysis which he employed was quite untrustworthy.

Thinking that by measuring the extent to which the *phlogistication* (we would now say the *deoxidation*) of common air was carried by mixing measured quantities of common and inflammable airs and exploding this mixture, he might be able to determine the amount of phlogiston in a given volume of inflammable air, he mixed the two airs in glass tubes, through the sides of which he had cemented two pieces of wire, sealed the tubes, and exploded the˙ mixture by passing electric sparks from wire to wire. The residual air now contained, according to Priestley, more phlogiston, and therefore relatively less dephlogisticated air than before the explosion. He made various measurements of the quantities of dephlogisticated air in the tubes, but without getting any constant results. He noticed that after the explosions the insides of the tubes were covered with moisture. At a later

time he exploded a mixture of dephlogisticated and inflammable airs (oxygen and hydrogen) in a copper globe, and recorded the fact that after the explosion the globe contained a little water. Priestley was here apparently on the eve of a great discovery. "In looking for one thing," says Priestley, " I have generally found another, and sometimes a thing of much more value than that which I was in quest of." Had he performed the experiment of exploding dephlogisticated and inflammable airs with more care, and had he made sure that the airs used were quite dry before the explosion, he would probably have found a thing of indeed much more value than that of which he was in quest ; he would probably have discovered the compound nature of water—a discovery which was made by Cavendish three or four years after these experiments described by Priestley.

Some very curious observations were made by Priestley regarding the colour of the gas obtained by heating "spirit of nitre" (*i.e.* nitric acid). He showed that a yellow gas or air is obtained by heating colourless liquid spirit of nitre in a sealed glass tube, and that as the heating is continued the colour of the gas gets darker, until it is finally very dark orange red. These experiments have found an explanation only in quite recent times.

Another discovery made by Priestley while in Birmingham, viz. that an acid is formed when electric sparks are passed through ordinary air for some time, led, in the hands of Cavendish—an experimenter who was as careful and deliberate as Priestley was rapid and careless—to the demonstration of the composition of nitric acid.

Many observations were made by Priestley on the effects of various airs on growing plants and living animals; indeed, one of his customary methods of testing different airs was to put a mouse into each and watch the effects of the air on its breathing. He grew sprigs of mint in common air, in dephlogisticated air (oxygen), and in phlogisticated air (nitrogen, but probably not pure); the sprig in the last-named air grew best, while that in the dephlogisticated air soon appeared sickly. He also showed that air which has been rendered "noxious" by the burning of a candle in it, or by respiration or putrefaction, could be restored to its original state by the action of growing plants. He thought that the air was in the first instance rendered noxious by being impregnated with phlogiston, and that the plant restored the air by removing this phlogiston. Thus Priestley distinctly showed that (to use his own words) "it is very probable that the injury which is continually done to the atmosphere by the respiration of such a number of animals as breathe it, and the putrefaction of such vast masses, both of vegetable and animal substances, exposed to it, is, in part at least, repaired by the vegetable creation." But from want of quantitative experiments he failed to give any just explanation of the process whereby this "reparation" is accomplished.

During his stay in Birmingham, Priestley was busily engaged, as was his wont during life, in writing metaphysical and theological treatises and pamphlets.

At this time the minds of men in England were much excited by the events of the French Revolution, then being enacted before them. Priestley and some of his friends were known to sympathize with the French people in this great struggle, as they had been on the side of the Americans in the War of Independence. Priestley's political opinions had, in fact, always been more advanced than the average opinion of his age ; by some he was regarded as a dangerous character. But if we read what he lays down as a fundamental proposition in the "Essay on the First Principles of Civil Government" (1768), we cannot surely find anything very startling.

" It must be understood, whether it be expressed or not, that all people live in society for their mutual advantage ; so that the good and happiness of the members, that is the majority of the members of any state, is the great standard by which everything relating to that state must be finally determined. And though it may be supposed that a body of people may be bound by a voluntary resignation of all their rights to a single person, or to a few, it can never be supposed that the resignation is obligatory on their posterity, because it is manifestly contrary to the good of the whole that it should be so."

Priestley proposed many political reforms, but he was decidedly of opinion that these ought to be brought about gradually. He was in favour of abolishing all religious State establishments, and was a declared enemy to the Church of England. His controversies with the clergy of Birmingham helped to stir up a section of public opinion against him, and to bring about the condemnation of his writings in many parts of the country ; he was also unfortunate in making an enemy of Mr. Burke, who spoke against him and his writings in the House of Commons.

In the year 1791, the day of the anniversary of the taking of the Bastille was celebrated by some of Priestley's friends in Birmingham. On that day a senseless mob, raising the cry of " Church and King," caused a riot in the town. Finding that they were not checked by those in authority, they after a time attacked and burned Dr. Priestley's meeting-house, and then destroyed his dwelling-house, and the houses of several other Dissenters in the town. One of his sons barely escaped with his life. He himself found it necessary to leave Birmingham for London, as he considered his life to be in danger. Many of his manuscripts, his library, and much of his apparatus were destroyed, and his house was burned.

A congregation at Hackney had the courage at this time to invite Priestley to become their minister. Here he remained for about three years, ministering to the congregation, and pursuing his

chemical and other experiments with the help of
apparatus and books which had been supplied by
his friends, and by the expenditure of part of the
sum, too small to cover his losses, given him by
Government in consideration of the damage done
to his property in the riots at Birmingham.

But finding himself more and more isolated and
lonely, especially after the departure of his three
sons to America, which occurred during these years,
he at last resolved to follow them, and spend
the remainder of his days in the New World.
Although Priestley had been very badly treated
by a considerable section of the English people,
yet he left his native country "without any resent-
ment or ill will." "When the time for reflection,"
he says, "shall come, my countrymen will, I am
confident, do me more justice." He left England
in 1795, and settled at Northumberland, in Penn-
sylvania, about a hundred and thirty miles north-
west of Philadelphia. By the help of his friends in
England he was enabled to build a house and
establish a laboratory and a library; an income
was also secured sufficient to maintain him in
moderate comfort.

The chair of chemistry in the University of
Philadelphia was offered to him, and he was also
invited to the charge of a Unitarian chapel in New
York; but he preferred to remain quietly at work
in his laboratory and library, rather than again to
enter into the noisy battle of life. In America
he published several writings. Of his chemical

discoveries made after leaving England, the most important was that an inflammable gas is obtained by heating metallic calces with carbon. The production of this gas was regarded by Priestley as an indisputable proof of the justness of the theory of phlogiston (see pp. 63, 64).

His health began to give way about 1801; gradually his strength declined, and in February 1804, the end came quietly and peacefully.

A list of the books and pamphlets published by Priestley on theological, metaphysical, philological, historical, educational and scientific subjects would fill several pages of this book. His industry was immense. To accomplish the vast amount of work which he did required the most careful outlay of time. In his "Memoirs," partly written by himself, he tells us that he inherited from his parents "a happy temperament of body and mind;" his father especially was always in good spirits, and "could have been happy in a workhouse." His paternal ancestors had, as a race, been healthy and long-lived. He was not himself robust as a youth, yet he was always able to study: "I have never found myself," he says, "less disposed or less qualified for mental exertion of any kind at one time of the day more than another; but all seasons have been equal to me, early or late, before dinner or after."

His peculiar evenness of disposition enabled him quickly to recover from the effects of any unpleasant occurrence; indeed, he assures us that "the most perfect satisfaction" often came a day or two after "an event that afflicted me the most, and without any change having taken place in the state of things."

Another circumstance which tended to make life easy to him was his fixed resolution, that in any controversy in which he might be engaged, he would frankly acknowledge every mistake he perceived himself to have fallen into.

Priestley's scientific work is marked by rapidity of execution. The different parts do not hang together well; we are presented with a brilliant series of discoveries, but we do not see the connecting strings of thought. We are not then astonished when he tells us that sometimes he forgot that he had made this or that experiment, and repeated what he had done weeks before. He says that he could not work in a hurry, and that he was therefore always methodical; but he adds that he sometimes blamed himself for "doing to-day what had better have been put off until to-morrow."

Many of his most startling discoveries were the results of chance operations, "not of themes worked out and applied." He was led to the discovery of oxygen, he says, by a succession of extraordinary accidents. But that he was able to take advantage of the chance observations, and from these to advance to definite facts, constitutes the essential difference between him and ordinary plodding investigators. Although he rarely, if ever, saw all the bearings of his own discoveries, although none of his experiments was accurately worked out to its conclusion, yet he did see, rapidly and as it appeared almost at one glance, something of their meanings, and this something was enough to urge him on to fresh experimental work.

Although we now condemn Priestley's theories as quite erroneous, yet we must admire his undaunted devotion to experiment. He was a true student of science in one essential point, viz. Nature was for him the first and the last court of appeal. He theorized and speculated much, he experimented rapidly and not accurately, but he was ever appealing to natural facts ; and in doing this he could not but lay some foundation which should remain. The facts discovered by him are amongst the very corner-stones on which the building of chemical science was afterwards raised.

So enthusiastic was Priestley in the prosecution of his experiments, that when he began, he tells us, " I spent all the money I could possibly raise, carried on by my ardour in philosophical investigation, and entirely regardless of consequences, except so far as never to contract any debts." He seems all through his life to have been perfectly free from anxiety about money affairs.

Priestley's manner of work shows how kindly and genial he was. He trained himself to talk and think and write with his family by the fireside ; " nothing but reading aloud, or speaking without interruption," was an obstruction to his work.

Priestley was just the man who was wanted in

the early days of chemical science. By the vast number, variety and novelty of his experimental results, he astonished scientific men—he forcibly drew attention to the science in which he laboured so hard; by the brilliancy of some of his experiments he obliged chemists to admit that a new field of research was opened before them, and the instruments for the prosecution of this research were placed in their hands; and even by the unsatisfactoriness of his reasoning he drew attention to the difficulties and contradictions of the theories which then prevailed in chemistry.

That the work of Priestley should bear full fruit it was necessary that a greater than he should interpret it, and should render definite that which Priestley had but vaguely shown to exist.

The man who did this, and who in doing it really established chemistry as a science, was Lavoisier.

But before considering the work of Lavoisier, I should like to point out that many of the physical characters of common air had been clearly established in the later years of the seventeenth century by the Honourable Robert Boyle. In the "Sceptical Chymist," published in 1661, Mr. Boyle had established the fact that air is a material substance possessed of weight, that this air presses on the surface of all things, and that by removing part of the air in an enclosed space the pressure within that space is diminished. He had demonstrated that the boiling point of water is dependent on the pressure of the air on the surface of the water. Having boiled some water " a pretty while, that by the heat it might be freed from the latitant air," he placed the vessel containing the hot water within the receiver of an arrangement which he had invented for sucking air out of an enclosed space ; as soon as he began to suck out air from this receiver, the water boiled "as if it had stood over a very quick fire. . . . Once, when the air had been drawn out, the liquor did, upon a single exsuction, boil so long with prodigiously vast bubbles, that the effervescence lasted almost as long as was requisite for the rehearsing of a *Pater noster.*" Boyle had gone further than the qualitative fact that the volume of an enclosed quantity of air alters with changes in the pressure to which that air is subjected; he had shown by simple and accurate experiments that "the volume varies inversely as the pressure." He had established the generalization of so much importance in physical science now known as *Boyle's law.*

The work of the Honourable Henry Cavendish will be considered in some detail in the book on "The Physicists" belonging to this series, but I must here briefly allude to the results of his experiments on air published in the *Philosophical Transactions* for 1784 and 1785.

Cavendish held the ordinary view that when a metal burns in air, the air is thereby phlogisticated ; but why is it, he asked, that the volume of air is decreased by this process ? It was very

generally said that fixed air was produced during
the calcination of metals, and was absorbed by the
calx. But Cavendish instituted a series of experiments which proved that no fixed air could be obtained from metallic calces. In 1766 inflammable
air (hydrogen) was discovered by Cavendish ; he
now proved that when this air is exploded with
dephlogisticated air (oxygen), water is produced.
He showed that when these two airs are mixed in
about the proportion of two volumes of hydrogen
to one volume of oxygen, the greater part, if not
the whole of the airs is condensed into water by
the action of the electric spark. He then proceeded to prove by experiments that when common
air is exploded with inflammable air water is likewise produced, and phlogisticated air (*i.e.* nitrogen)
remains.

Priestley and Cavendish had thus distinctly
established the existence of three kinds of air,
viz. dephlogisticated air, phlogisticated air, and
inflammable air. Cavendish had shown that
when the last named is exploded with common
air water is produced (which is composed of
dephlogisticated and inflammable airs), and phlogisticated air remains. Common air had thus
been proved to consist of these two—phlogisticated
and dephlogisticated airs (nitrogen and oxygen).
Applying these results to the phenomenon of the
calcination of metals, Cavendish gave reasons for
thinking that the metals act towards common air
in a manner analogous to that in which inflam-

mable air acts—that they withdraw dephlogisticated and leave phlogisticated air ; but, as he was
a supporter of the phlogistic theory, he rather
preferred to say that the burning metals withdraw
dephlogisticated air and phlogisticate that which
remains ; in other words, while admitting that a
metal in the process of burning gains dephlogisticated air, he still thought that the metal also loses
something, viz. phlogiston.

That Cavendish in 1783–84 had proved air to
consist of two distinct gases, and water to be produced by the union of two gases, must be remembered as we proceed with the story of the discoveries
of Lavoisier.

ANTOINE LAURENT LAVOISIER, born in Paris in
1743, was the son of a wealthy merchant, who,
judging from his friendship with many of the men
of science of that day, was probably of a scientific
bent of mind, and who certainly showed that he
was a man of sense by giving his son the best
education which he could obtain. After studying
in the Mazarin College, Lavoisier entered on a
course of training in physical, astronomical, botanical and chemical science. The effects of this
training in the accurate methods of physics are
apparent in the chemical researches of Lavoisier.

At the age of twenty-one Lavoisier wrote a
memoir which gained the prize offered by the
French Government for the best and most econo-
mical method of lighting the streets of a large city.

While making experiments, the results of which were detailed in this paper, Lavoisier lived for six weeks in rooms lighted only by artificial light, in order that his eyesight might become accustomed to small differences in the intensities of light from various sources. When he was twenty-five years old Lavoisier was elected a member of the Academy of Sciences. During the next six years (1768–1774) he published various papers, some on chemical, some on geological, and some on mathematical subjects. Indeed at this time, although an ardent cultivator of natural science, he appears to have been undecided as to which branch of science he should devote his strength.

The accuracy and thoroughness of Lavoisier's work, and the acuteness of his reasoning powers, are admirably illustrated in two papers, published in the Memoirs of the Academy for 1770, on the alleged conversion of water into earth.

When water is boiled for a long time in a glass vessel a considerable quantity of white siliceous earth is found in the vessel. This apparent conversion or transmutation of water into earthy matter was quite in keeping with the doctrines which had been handed down from the times of the alchemists; the experiment was generally regarded as conclusively proving the possibility of changing water into earth. Lavoisier found that after heating water for a hundred and one days in a closed *and weighed* glass vessel, there was no change in the total weight of the vessel and its contents; when he poured out the water and evaporated it to dryness, he obtained 20·4 grains of solid earthy matter; but he also found, what had been before overlooked, that the glass vessel had lost weight. The actual loss amounted to 17·4 grains. The difference between this and the weight of the earthy matter in the water, viz. three grains, was set down (and as we now know justly set down) by Lavoisier to errors of experiment. Lavoisier therefore concluded that water, when boiled, is not changed into earth, but that a portion of the earthy matter of which glass is composed is dissolved by the water. This conclusion was afterwards confirmed by the Swedish chemist *Scheele*, who proved that the composition of the earthy matter found in the water is identical with that of some of the constituents of glass.

By this experiment Lavoisier proved the old alchemical notion of transmutation to be erroneous; he showed that water is not transmuted into earth, but that each of these substances is possessed of definite properties which belong to it and to it only. He established the all-important generalization—which subsequent research has more amply confirmed, until it is to-day accepted as the very foundation of every branch of physical science—that in no process of change is there any alteration in the total mass of matter taking part in that change. The glass vessel in which Lavoisier boiled water for so many days lost weight; but the matter lost by the glass was found dissolved in the water.

We know that this generalization holds good in all chemical changes. Solid sulphur may be converted into liquid oil of vitriol, but it is only by the sulphur combining with other kinds of matter ; the weight of oil of vitriol produced is always exactly equal to the sum of the weights of the sulphur, hydrogen and oxygen which have combined to form it. The colourless gases, hydrogen and oxygen, combine, and the limpid liquid water is the result ; but the weight of the water produced is equal to the sum of the weights of hydrogen and oxygen which combined together. It is impossible to overrate the importance of the principle of the *conservation of mass*, first definitely established by Lavoisier.

Some time about the year 1770 Lavoisier turned his attention seriously to chemical phenomena. In 1774 he published a volume entitled " Essays Physical and Chemical," wherein he gave an historical account of all that had been done on the subject of airs from the time of Paracelsus to the year 1774, and added an account of his own experiments, in which he had established the facts that a metal in burning absorbs air, and that when the metallic calx is reduced to metal by heating with charcoal, an air is produced of the same nature as the fixed air of Dr. Black.

In November 1772 Lavoisier deposited a sealed note in the hands of the Secretary to the Academy of Sciences. This note was opened on the 1st of May 1773, and found to run as follows * :—

"About eight days ago I discovered that sulphur in burning, far from losing, augments in weight; that is to say, that from one pound of sulphur much more than one pound of vitriolic acid is obtained, without reckoning the humidity of the air. Phosphorus presents the same phenomenon. This augmentation of weight arises from a great quantity of air which becomes fixed during the combustion, and which combines with the vapours.

"This discovery, confirmed by experiments which I regard as decisive, led me to think that what is observed in the combustion of sulphur and phosphorus might likewise take place with respect to all the bodies which augment in weight by combustion and calcination ; and I was persuaded that the augmentation of weight in the calces of metals proceeded from the same cause. The experiment fully confirmed my conjectures.

"I operated the reduction of litharge in closed vessels with Hale's apparatus, and I observed that at the moment of the passage of the calx into the metallic state, there was a disengagement of air in considerable quantity, and that this air formed a volume at least one thousand times greater than that of the litharge employed.

"As this discovery appears to me one of the most interesting which has been made since Stahl, I

* The translation is taken from Thomson's "History of Chemistry."

thought it expedient to secure to myself the property, by depositing the present note in the hands of the Secretary of the Academy, to remain secret till the period when I shall publish my experiments.

"LAVOISIER.

"Paris, 11th November 1772."

In his paper "On the Calcination of Tin in Closed Vessels, and on the Cause of Increase of Weight acquired by the Metal during this Process" (published in 1774), we see and admire Lavoisier's manner of working. A weighed quantity (about half a pound) of tin was heated to melting in a glass retort, the beak of which was drawn out to a very small opening; the air within the retort having expanded, the opening was closed by melting the glass before the blowpipe. The weight of retort and tin was now noted; the tin was again heated to its melting point, and kept at this temperature as long as the process of calcination appeared to proceed; the retort and its contents were then allowed to cool and again weighed. No change was caused by the heating process in the total weight of the whole apparatus. The end of the retort beak was now broken off; air rushed in with a hissing sound. The retort and contents were again weighed, and the increase over the weight at the moment of sealing the retort was noted. The calcined tin in the retort was now collected and weighed. It was found that the increase in the weight of the tin was equal to the weight of the air which rushed into the retort. Hence Lavoisier concluded that the calcination of tin was accompanied by an absorption of air, and that the difference between the weights of the tin and the calx of tin was equal to the weight of air absorbed; but he states that probably only a part of the air had combined with the tin, and that hence air is not a simple substance, but is composed of two or more constituents.

Between the date of this publication and that of Lavoisier's next paper on combustion we know that Priestley visited Paris. In his last work, "The Doctrine of Phlogiston established" (published in 1800), Priestley says, "Having made the discovery of dephlogisticated air some time before I was in Paris in 1774, I mentioned it at the table of Mr. Lavoisier, when most of the philosophical people in the city were present; saying that it was a kind of air in which a candle burned much better than in common air, but I had not then given it any name. At this all the company, and Mr. and Mrs. Lavoisier as much as any, expressed great surprise. I told them that I had got it from *precipitatum per se*, and also from *red lead*."

In 1775 Lavoisier's paper, "On the Nature of the Principle which combines with the Metals during their Calcination, and which augments their Weight," was read before the Academy. The preparation and properties of an air obtained, in November 1774, from *red precipitate* are described, but Priestley's name is not mentioned. It seems probable, however, that Lavoisier learned the existence and the mode of preparation of this air from Priestley;* but we have seen that even in 1779 Priestley was quite in the dark as to the true nature of the air discovered by him (p. 60).

In papers published in the next three or four years Lavoisier gradually defined and more thoroughly explained the phenomenon of combustion. He burned phosphorus in a confined volume of air, and found that about one-fourth of the air disappeared, that the residual portion of air was unable to support combustion or to sustain animal life, that the phosphorus was converted into a white substance deposited on the sides of the vessel in which the experiment was performed, and that for each grain of phosphorus used about two and a half grains of this white solid were obtained. He further described the properties of the substance produced by burning phosphorus, gave it the name of *phosphoric acid*, and described some of the substances formed by combining it with various bases.

The burning of candles in air was about this time studied by Lavoisier. He regarded his experiments as proving that the air which remained after burning a candle, and in which animal life could not be sustained, was really present before the burning; that common air consisted of about one-fourth part of dephlogisticated air and three-

* Nevertheless, in other places Lavoisier most readily acknowledges the merits of Priestley.

fourths of *azotic air* (*i.e.* air incapable of sustaining life) ; and that the burning candle simply combined with, and so removed the former of these, and at the same time produced more or less fixed air.

In his treatise on chemistry Lavoisier describes more fully his proof that the calcination of a metal consists in the removal, by the metal, of dephlogisticated air (or oxygen) from the atmosphere, and that the metallic calx is simply a compound of metal and oxygen. The experiments are strictly quantitative and are thoroughly conclusive. He placed four ounces of pure mercury in a glass balloon, the neck of which dipped beneath the surface of mercury in a glass dish, and then passed a little way up into a jar containing fifty cubic inches of air, and standing in the mercury in the dish. There was thus free communication between the air in the balloon and that in the glass jar, but no communication between the air inside and that outside the whole apparatus. The mercury in the balloon was heated nearly to its boiling point for twelve days, during which time red-coloured specks gradually formed on the surface of the metal ; at the end of this time it was found that the air in the glass jar measured between forty-two and forty-three cubic inches. The red specks when collected amounted to forty-five grains ; they were heated in a very small retort connected with a graduated glass cylinder containing mercury. Between seven and eight cubic inches of pure dephlogisticated air (oxygen) were obtained in this cylinder, and forty-one and a half grains of metallic mercury remained when the decomposition of the red substance was completed.

The conclusion drawn by Lavoisier from these experiments was that mercury, when heated nearly to boiling in contact with air, withdraws oxygen from the air and combines with this gas to form *red precipitate*, and that when the red precipitate which has been thus formed is strongly heated, it parts with the whole of its oxygen, and is changed back again into metallic mercury.

Lavoisier had now (1777–78) proved that the calces of mercury, tin and lead are compounds of these metals with oxygen ; and that the oxygen is obtained from the atmosphere when the metal burns. But the phlogistic chemistry was not yet overthrown. We have seen that the upholders of phlogiston believed that in the inflammable air of Cavendish they had at last succeeded in obtaining the long-sought-for phlogiston. Now they triumphantly asked, Why, when metals dissolve in diluted vitriolic or muriatic acid with evolution of inflammable air, are calces of these metals produced ? And they answered as triumphantly, Because these metals lose phlogiston by this process, and we know that a calx is a metal deprived of its phlogiston.

Lavoisier contented himself with observing that a metallic calx always weighed more than the metal from which it was produced ; and that as inflammable air, although much lighter than common air, was distinctly possessed of weight, it was not possible that a metallic calx could be metal deprived of inflammable air. He had given a simple explanation of the process of calcination, and had proved, by accurate experiments, that this explanation was certainly true in some cases. Although all the known facts about solution of metals in acids could not as yet be brought within his explanation, yet none of these facts was absolutely contradictory of that explanation. He was content to wait for further knowledge. And to gain this further knowledge he set about devising and performing new experiments. The upholders of the theory of phlogiston laid considerable stress on the fact that metals are produced by heating metallic calces in inflammable air; the air is absorbed, they said, and so the metal is reproduced. It was obviously of the utmost importance that Lavoisier should learn more about this inflammable air, and especially that he should know exactly what happened when this air was burned. He therefore prepared to burn a large quantity of inflammable air, arranging the experiment so that he should be able to collect and examine the product of this burning, whatever should be the nature of that product. But at this time the news was brought to Paris that Cavendish had obtained water by burning mixtures of inflammable and dephlogisticated airs. This must have been a most exciting announcement to

Lavoisier; he saw how much depended on the accuracy of this statement, and as a true student of Nature, he at once set about to prove or disprove it. On the 24th of June 1783, in the presence of the King and several notabilities (including Sir Charles Blagden, Secretary of the Royal Society, who had told Lavoisier of the experiments of Cavendish), Lavoisier and Laplace burned inflammable and dephlogisticated airs, and obtained water. As the result of these experiments they determined that one volume of dephlogisticated air combines with 1·91 volumes of inflammable air to form water.

A little later Lavoisier completed the proof of the composition of water by showing that when steam is passed through a tube containing iron filings kept red hot, inflammable air is evolved and calx of iron remains in the tube.

Lavoisier could now explain the conversion of a metallic calx into metal by the action of inflammable air; this air decomposes the calx—that is, the metallic oxide—combines with its oxygen to form water, and so the metal is produced.

When a metal is dissolved in diluted vitriolic or muriatic acid a calx is formed, because, according to Lavoisier, the water present is decomposed by the metal, inflammable air is evolved, and the dephlogisticated air of the water combines with the metal forming a calx, which then dissolves in the acid.

Lavoisier now studied the properties of the compounds produced by burning phosphorus, sulphur and carbon in dephlogisticated air. He found that solutions of these compounds in water had a more or less sour taste and turned certain blue colouring matters red ; but these were the properties regarded as especially belonging to acids. These products of combustion in dephlogisticated air were therefore acids ; but as phosphorus, carbon and sulphur were not themselves acids, the acid character of the substances obtained by burning these bodies in dephlogisticated air must be due to the presence in them of this air. Hence Lavoisier concluded that this air is the substance the presence of which in a compound confers acid properties on that compound. This view of the action of dephlogisticated air he perpetuated in the name "oxygen" (from Greek, = *acid-producer*), which he gave to dephlogisticated air, and by which name this gas has ever since been known.

Priestley was of opinion that the atmosphere is rendered noxious by the breathing of animals, because it is thereby much phlogisticated, and he thought that his experiments rendered it very probable that plants are able to purify this noxious air by taking away phlogiston from it (see p. 69). But Lavoisier was now able to give a much more definite account of the effects on the atmosphere of animal and vegetable life. He had already shown that ordinary air contains oxygen and azote (nitrogen), and that the former is alone concerned in the process of combustion. He was now able to show that animals during respiration draw in air into their lungs ; that

a portion of the oxygen is there combined with
carbon to form carbonic acid gas (as the fixed air
of Black was now generally called), which is again
expired along with unaltered azote. Respiration
was thus proved to be a process chemically analogous to that of calcination.

Thus, about the year 1784–85, the theory of
phlogiston appeared to be quite overthrown. The
arguments of its upholders, after this time, were
not founded on facts; they consisted of fanciful
interpretations of crudely performed experiments.
Cavendish was the only opponent to be dreaded
by the supporters of the new chemistry. But we
have seen that although Cavendish retained the
language of the phlogistic theory (see pp. 78, 79) as
in his opinion equally applicable to the facts of combustion with that of the new or Lavoisierian theory,
he nevertheless practically admitted the essential
point of the latter, viz. that calces are compounds
of metal and oxygen (or dephlogisticated air).
Although Cavendish was the first to show that
water is produced when the two gases hydrogen
and oxygen are exploded together, it would yet
appear that he did not fully grasp the fact that
water is a compound of these two gases; it was
left to Lavoisier to give a clear statement of this
all-important fact, and thus to remove the last prop
from under the now tottering, but once stately
edifice built by Stahl and his successors.

The explanation given by Lavoisier of combustion was to a great extent based on a concep-

tion of element and compound very different from that of the older chemists.   In the "Sceptical Chymist" (1661) Boyle had argued strongly against the doctrine of the four "elementary principles," earth, air, fire and water, as held by the "vulgar chymists."   The existence of these principles, or some of them, in every compound substance was firmly held by most chemists in Boyle's time.   They argued thus : when a piece of green wood burns, the existence in the wood of the principle of fire is made evident by the flame, of the principle of air by the smoke which ascends, of that of water by the hissing and boiling sound, and of the principle of earth by the ashes which remain when the burning is finished.*

Boyle combated the inference that because a flame is visible round the burning wood, and a light air or smoke ascends from it, *therefore* these principles were contained in the wood before combustion began.   He tried to prove by experiments that one substance may be obtained from another in which the first substance did not already exist ; thus, he heated water for a year in a closed glass vessel, and obtained solid particles heavier than, and as he supposed formed from, the water.   We have already

---

* A similar method of reasoning was employed so far back as the tenth century : thus, in an Anglo-Saxon "Manual of Astronomy" we read, "There is no corporeal thing which has not in it the four elements, that is, air and fire, earth and water. . . . Take a stick and rub it on something, it becomes hot directly with the fire which lurks in it ; burn one end, then goeth the moisture out at the other end with the smoke."

learned the true interpretation of this experiment from the work of Lavoisier. Boyle grew various vegetables in water only, and thought that he had thus changed water into solid vegetable matter. He tells travellers' tales of the growth of pieces of iron and other metals in the earth or while kept in underground cellars.

We now know how erroneous in most points this reasoning was, but we must admit that Boyle established one point most satisfactorily, viz. that because earth, or air, or fire, or water is obtained by heating or otherwise decomposing a substance, it does not necessarily follow that the earth, or air, or fire, or water existed as such in the original substance. He overthrew the doctrine of elementary principles held by the "vulgar chymists." Defining elements as "certain primitive and simple bodies which, not being made of any other bodies, or of one another, are the ingredients of which all those called perfectly mixt bodies are immediately compounded, and into which they are ultimately resolved," Boyle admitted the *possible* existence, but thought that the facts known at his time did not warrant the assertion of the *certain* existence, of such "elements." The work of Hooke and Mayow on combustion tended to strengthen this definition of "element" given by Boyle.

Black, as we have seen, clearly proved that certain chemical substances were possessed of definite and unvarying composition and properties ; and Lavoisier, indirectly by his explanation of combustion, and directly in his "Treatise on Chemistry," laid down the definition of "element" which is now universally adopted.

An element is a substance from which no simpler forms of matter—that is, no forms of matter each weighing less than the original substance—have *as yet* been obtained.

In the decade 1774–1784 chemical science was thus established on a sure foundation by Lavoisier. Like most great builders, whether of physical or mental structures, he used the materials gathered by those who came before him, but the merit of arranging these materials into a well-laid foundation, on which the future building might firmly rest, is due to him alone.

The value of Lavoisier's work now began to be recognized by his fellow-chemists in France. In 1785 Berthollet, one of the most rising of the younger French chemists, declared himself a convert to the views of Lavoisier on combustion. Fourcroy, another member of the Academy, soon followed the example of Berthollet. Fourcroy, knowing the weakness of his countrymen, saw that if the new views could be made to appear as especially the views of Frenchmen, the victory would be won; he therefore gave to the theory of Lavoisier the name "*La chimie Française.*" Although this name was obviously unfair to Lavoisier, it nevertheless caused the antiphlogistic theory to be identified with the French chemists, and succeeded in impressing the French public generally with the idea that to hold to the old theory was to be a traitor to the glory of one's country. M. de Morveau, who held a prominent place both in politics and science, was invited to Paris, and before long was persuaded to embrace the new theory. This conversion—for " the whole matter was managed as if it had been a political intrigue rather than a philosophical inquiry "—was of great importance to Lavoisier and his friends. M. de Morveau was editor of the chemical part of the " Encyclopédie Méthodique ; " in that part of this work which had appeared before 1784 De Morveau had skilfully opposed the opinions of Lavoisier, but in the second part of the work he introduced an advertisement announcing the change in his opinions on the subject of combustion, and giving his reasons for this change.

The importance of having a definite language in every science is apparent at each step of advance. Lavoisier found great difficulty in making his opinions clear because he was obliged to use a language which had been introduced by the phlogistic chemists, and which bore the impress of that theory on most of its terms. About the years 1785–1787, Lavoisier, Berthollet, Fourcroy and De Morveau drew up a new system of chemical nomenclature. The fundamental principles of that system have remained as those of every nomenclature since proposed. They are briefly these :—

An element is a substance from which no form of matter simpler than itself has as yet been obtained.

Every substance is to be regarded as an element until it is proved to be otherwise.

The name of every compound is to tell of what elements the substance is composed, and it is to express as far as possible the relative amounts of the elements which go to form the compound.

Thus the compounds of oxygen with any other element were called oxides, *e.g.* iron oxide, mercury oxide, tin oxide, etc. When two oxides of iron came to be known, one containing more oxygen relatively to the amount of iron present than the other, that with the greater quantity of oxygen was called iron peroxide, and that with the smaller quantity iron protoxide.

We now generally prefer to use the name of the element other than oxygen in adjectival form, and to indicate the relatively smaller or greater quantity of oxygen present by modifications in the termination of this adjective. Thus iron protoxide is now generally known as ferr*ous* oxide, and iron peroxide as ferr*ic* oxide. But the principles laid down by the four French chemists in 1785–1787 remain as the groundwork of our present system of nomenclature.

The antiphlogistic theory was soon adopted by all French chemists of note. We have already seen that Black, with his usual candour and openness to conviction, adopted and taught this theory, and we are assured by Dr. Thomas Thomson that when he attended Black's classes, nine years after the publication of the French system of nomenclature, that

III.   II

system was in general use among the chemical students of the university. The older theory was naturally upheld by the countrymen of the distinguished Stahl after it had been given up in France. In the year 1792 Klaproth, who was then Professor of Chemistry in Berlin, proposed to the Berlin Academy of Sciences to repeat the more important experiments on which the Lavoisierian theory rested, before the Academy. His offer was accepted, and from that time most of the Berlin chemists declared themselves in favour of the new theory.

By the close of last century the teaching or Lavoisier regarding combustion found almost universal assent among chemists. But this teaching carried with it, as necessary parts, the fundamental distinction between element and compound ; the denial of the existence of "principles" or "essences ;" the recognition of the study of actually occurring reactions between substances as the basis on which all true chemical knowledge was to be built; and the full acknowledgment of the fact that matter is neither created nor destroyed, but only changed as to its form, in any chemical reaction.

Of Lavoisier's other work I can only mention the paper on "Specific Heats" contributed by Laplace and Lavoisier to the Memoirs of the Academy for 1780. In this paper is described the ice calorimeter, whereby the amount of heat given out by a substance in cooling from one definite temperature to another is determined, by measuring the amount of ice converted into water by the heated substance in cooling through the stated interval of temperature. The specific heats of various substances, *e.g.* iron, glass, mercury, quicklime, etc., were determined by the help of this instrument.

As we read the record of work done by Lavoisier during the years between 1774 and 1794—work which must have involved a great amount of concentrated thought as well as the expenditure of much time—we find it hard to realize that the most tremendous political and social revolution which the modern world has seen was raging around him during this time.

In the earlier days of the French Revolution, and in the time immediately preceding that movement, many minds had been stirred to see the importance of the study of Nature ; but it was impossible that natural science should continue to flourish when the tyrant Robespierre had begun the Reign of Terror.

The roll of those who perished during this time contains no more illustrious name than that of Antoine Laurent Lavoisier. In the year 1794 Lavoisier, who had for some time acted as a *fermier-général* under the Government, was accused of mixing with the tobacco "water and other ingredients hurtful to the health of the citizens." On this pretext he and some of his colleagues were condemned to death. For some days Lavoisier found a hiding-place among his friends, but hearing that his colleagues had been arrested, he delivered himself up to the authorities, only asking that the death sentence should not be executed until he had completed the research in which he was engaged; "not" that he was "unwilling to part with life," but because he thought the results would be "for the good of humanity."

"The Republic has no need of chemists; the course of justice cannot be suspended," was the reply.

On the 8th of May 1794, the guillotine did its work; and in his fifty-first year Lavoisier "joined the majority." To the honour of the Academy of which he was so illustrious a member it is recorded that a deputation of his fellow-workers in science, braving the wrath of Robespierre, penetrated to the dungeons of the prison and placed a wreath on the grave of their comrade.

The period of the infancy of chemical science which I have now briefly described is broadly contemporaneous with the second half of the eighteenth century.

At this time the minds of men were greatly stirred. Opinions and beliefs consecrated by the assent of generations of men were questioned or denied; the pretensions of civil and ecclesiastical authorities were withstood; assertions however strongly made, and by whatever authority supported, were met by demands for reasons. In France this revolt against mere authority was especially marked. Led by the great thinker

Voltaire, the French philosophers attacked the generally accepted views in moral, theological and historical matters. A little later they began to turn with eager attention and hope to the facts of external Nature. Physical science was cultivated with wonderful vigour and with surprising success.

In the sciences of heat and light we have at this time the all-important works of Fourier, Prevost and Fresnel ; in geology and natural history we have Buffon and Cuvier ; the name of Bichat marks the beginning of biological science, and chemistry takes rank as a science only from the time of Lavoisier.

From the philosophers an interest in natural science spread through the mass of the people. About the year 1870 the lecture-rooms of the great teachers of chemistry, astronomy, electricity, and even anatomy were crowded with ladies and gentlemen of fashion in the French capital. A similar state of matters was noticeable in this country. Dr. Black's lecture theatre was filled by an audience which comprised many young men of good position. To know something of chemistry became an essential part of the training of all who desired to be liberally educated.

The secrets of Nature were now rapidly explored ; astonishing advances were made, and as a matter of course much opposition was raised.

In this active, inquiring atmosphere the young science of chemistry grew towards maturity.

Priestley, ever seeking for new facts, announcing

discovery after discovery, attacking popular belief
in most matters, yet satisfied to interpret his scientific discoveries in terms of the hypothesis with
which he was most familiar, was the pioneer of the
advancing science. He may be compared to the
advance-guard sent forward by the explorers of a
new country with orders to clear a way for the
main body : his work was not to level the rough
parts of the way, or to fill in the miry places with
well-laid metal, but rather rapidly to make a road
as far into the heart of the country as possible.

And we have seen how well he did the work.
In his discovery of various kinds of airs, notably
of oxygen, he laid the basis of the great generalizations of Lavoisier, and, what was perhaps of even
more importance, he introduced a new method into
chemistry. He showed the existence of a new and
unexplored region. Before his time, Hooke and
Mayow had proved the existence of more than one
kind of air, but the chemistry of gases arose with
the discoveries of Priestley.

Although Black's chief research, on fixed air and
on latent heat, was completed fifteen or twenty
years before Priestley's discovery of oxygen, yet
the kind of work done by Black, and its influence
on chemical science, mark him as coming after
Priestley in order of development. We have seen
that the work of Black was characterized by
thoroughness and suggestiveness. The largeness
of scope, the breadth of view, of this great philosopher are best illustrated in his discourses on

heat ; he there leads us with him in his survey of the domain of Nature, and although he tells us that hypotheses are a "mere waste of time," we find that it is by the strength of his imagination that he commands assent. But he never allows the imagination to degenerate into fanciful guesses ; he vigorously tests the fundamental facts of his theory, and then he uses the imagination in developing the necessary consequences of these facts.

To Black we owe not only the first rigorously accurate chemical investigation, but also the establishment of just ideas concerning the nature of heat.

But Lavoisier came before us as a greater than either Priestley or Black. To great accuracy and great breadth of view he added wonderful power of generalizing; with these, aided by marked mental activity and, on the whole, favourable external circumstances, he was able finally to overthrow the loose opinions regarding combustion and elementary principles which prevailed before his time, and so to establish chemistry as one of the natural sciences.

At the close of the first period of advance we find that the sphere of chemistry has been defined ; that the object of the science has been laid down, as being to find an explanation of the remarkable changes noticed in the properties of bodies ; that as a first step towards the wished-for explanation, all material substances have been divided by the chemist into elements and compounds ; that an element has been defined as any kind of matter from a given weight of which no simpler forms of matter—that is, no kinds of matter each weighing less than the original matter—have as yet been obtained; that the great principle of the indestructibility of matter has been established, viz. that however the properties of matter may be altered, yet the total mass (or quantity) remains unchanged; and lastly, we find that an explanation of one important class of chemical changes—those changes which occur when substances burn—has been found.

And we have also learned that the method by which these results were obtained was this—to go to Nature, to observe and experiment accurately, to consider carefully the results of these experiments, and so to form a general hypothesis; by the use of the mental powers, and notably by the use of the imagination, to develop the necessary deductions from this hypothesis; and finally, to try these deductions by again inquiring from Nature "whether these things were so."

Before the time which we have been considering the paths of chemical science had scarcely yet been trodden. Each discovery was full of promise, each advance displayed the possibility of further progress; the atmosphere was filled as with "a mighty rushing wind" ready to sweep away the old order of things. The age was an age of doubt and of freedom from the trammels of authority; it was a time eminently suited for making advances in natural knowledge.

In the unceasing activity of Priestley and Lavoisier we may trace the influence of the restlessness of the age; but in the quietness and strength of the best work of these men, and notably in the work of Black; in the calmness with which Priestley bore his misfortunes at Birmingham; in the noble words of Lavoisier, "I am not unwilling to part with life, but I ask time to finish my experiments, because the results will, I believe, be for the good of humanity"—we see the truth of the assertion made by one who was himself a faithful student of Nature—

> "Nature never did betray
> The heart that loved her."

CHAPTER III.

ESTABLISHMENT OF GENERAL PRINCIPLES OF CHEMICAL SCIENCE—PERIOD OF DALTON.

John Dalton, 1766-1844.

THE progress of chemical knowledge became so rapid in the early years of the present century, that although I have in this chapter called the time immediately succeeding that of Lavoisier "the period of John Dalton," and although I shall attempt to describe the advances made by this philosopher without considering those of his contemporaries Davy and Berzelius, yet I must insist on the facts that this arrangement is made purely for the sake of convenience, and that many of the discoveries of Davy, Berzelius and others came in order of time before, or followed close upon the publication of Dalton's atomic theory.

Nevertheless, as the work of these men belongs in its essence to the modern period, and as the promulgation of the atomic theory by Dalton marks the beginning of this period, it seems better that we should have a clear conception of what was done by this chemist before proceeding to consider the advances made by his contemporaries and successors.

JOHN DALTON, the second of three children of Joseph and Deborah Dalton, was born at Eaglesfield, a village near Cockermouth, in Cumberland, on the 5th of September 1766. One of the first meeting-houses established by the Society of Friends is to be found in Eaglesfield.

The Dalton family had been settled for several generations on a small copyhold estate in this village. The first of them to join the Friends was the grandfather of John Dalton; his descendants remained faithful adherents of this society.

Dalton attended the village schools of Eaglesfield and the neighbourhood until he was eleven years old, by which time, in addition to learning reading, writing and arithmetic, he had "gone through a course of mensuration, surveying, navigation, etc." At the age of ten his taste for measurements and calculations began to be remarked by those around him; this taste was encouraged by Mr. Robinson, a relative of Dalton, who recognizing the indomitable perseverance of the boy appears to have taken some care about this time in directing his mathematical studies.

At the early age of twelve Dalton affixed to the door of his father's house a large sheet of paper whereon he announced that he had opened a school for youth of both sexes; also that "paper, pens and ink" were sold within. The boy-teacher had little authority over his pupils, who challenged their master to fight in the graveyard, and broke the windows of the room into which they had been locked till their tasks should be learned.

When he was fifteen years old Dalton removed to Kendal, where he continued for eleven or twelve years, at first as assistant-master, and then, along with his elder brother Jonathan, as principal of a boarding school for boys.

It was announced by the brothers that in this school "youth will be carefully instructed in English, Latin, Greek and French; also writing, arithmetic, merchants' accounts and the mathematics." The school was not very successful. Both brothers were hard, inflexible, and ungainly in their habits, and neither was fitted to become a successful teacher of boys: of the two, John had the gentler disposition, and was preferred by the boys; "besides, his mind was so occupied by mathematics that their faults escaped his notice."

During this time Dalton employed his leisure in learning Latin, Greek and French, and in pursuing his studies in mathematics and natural philosophy. He became a frequent contributor to the *Gentlemen's Diary*, a paper which received problems of various kinds—chiefly mathematical—and presented prizes for their successful solution.

Besides setting and answering mathematical

problems in this journal, and also in the *Ladies' Diary*, Dalton sometimes ventured into the wider fields of mental phenomena.   It seems strange to read that, even at the age of twenty-six, Dalton should occupy his leisure time composing answers to such queries as these :—

"Whether, to a generous mind, is the conferring or receiving an obligation, the greater pleasure ?"

"Is it possible for a person of sensibility and virtue, who has once felt the passion of love in the fullest extent that the human heart is capable of receiving it (being by death or some other circumstance for ever deprived of the object of its wishes), ever to feel an equal passion for any other object ?"

In his answer to the second of these queries, Dalton carefully framed two hypotheses, and as carefully drew conclusions from each.   The question in the *Diary* was by "Mira ;" if "Mira" were a "rapturous maiden" she would not derive much comfort from the cold and mathematical answer by "Mr. John Dalton of Kendal."

At Kendal Dalton made the acquaintance of Mr. Gough, who was about eight years older than Dalton, and had been blind from the age of two.   Mr. Gough, we are assured by Dalton, was "a perfect master of the Latin, Greek and French tongues ;" he understood "well all the different branches of mathematics ;" there was "no branch of natural philosophy but what he was well acquainted with ;" he knew "by the touch, taste and smell, almost every plant within twenty miles of Kendal." To the friendship of this remarkable man Dalton owed much ; with his help he acquired a fair knowledge of the classical languages, and he it was who set Dalton the example of keeping a regular record of weather observations.

On the 24th of March 1787 Dalton made his first entry in a book which he entitled " Observations on the Weather, etc. ; " the last entry in this book he made fifty-seven years later on the evening preceding his death. The importance of Dalton's meteorological observations, as leading him to the conception of the atomic theory, will be noticed as we proceed.

In the year 1793 Dalton, who was now twenty-seven years of age, was invited to Manchester to become tutor in the mathematical and natural philosophy department of a college recently established by influential Dissenters in that town. Eighty pounds for the session of ten months was guaranteed him ; and he was provided with "rooms and commons " in the college at a charge of £27 10s. per session.

He held this appointment for six years, when he retired, and continuing to live in Manchester devoted himself to researches in natural philosophy, gaining a living by giving private lessons in mathematics and physical science at a charge of 2s. 6d. per hour, or 1s. 6d. each if more than two pupils attended at the same time.

Dalton was elected a Fellow of the Literary

and Philosophical Society of Manchester in the year 1794; and from the time of his retiring from the tutorship of Manchester New College till the close of his life he spent a great part of his time in a room in the society's house in George Street, in studying and teaching. The fifty years thus spent are marked by few outward events. The history of Dalton's life from this time is the history of the development of his intellect, and the record of his scientific discoveries.

On one occasion during Dalton's stay at Kendal, as he was about to make a visit to his native village, he bethought himself that the present of a pair of silken hose would be acceptable to his mother. He accordingly purchased a pair marked "newest fashion;" but his mother's remark, "Thou hast brought me a pair of grand hose, John; but what made thee fancy so light a colour? I can never show myself at meeting in them," rather disconcerted him, as to his eyes the hose were of the orthodox drab colour. His mother insisted that the stockings were "as red as a cherry." John's brother upheld the "drab" side of the dispute; so the neighbours were called in, and gave their decision that the hose were "varra fine stuff, but uncommon scarlety."

From this time Dalton made observations on the peculiarities of his own vision and that of others, and in his first paper read before the Literary and Philosophical Society in 1794, he described these peculiarities. He says, "Since the year 1790 the occasional study of botany obliged me to attend more to colour than before. With respect to colours that were white, yellow, or green, I readily assented to the appropriate term ; blue, purple, pink and crimson appeared rather less distinguishable, being, according to my idea, all referable to blue. I have often seriously asked a person whether a flower was blue or pink, but was generally considered to be in jest." Dalton's colourblindness was amusingly illustrated at a later time, when having been created D.C.L. by the University of Oxford he continued to wear the red robes of his degree for some days ; and when his attention was drawn to the somewhat strange phenomenon, even in a university town, of an elderly gentleman in the dress of a Quaker perambulating the town day after day in a scarlet robe, he remarked that to him the gown appeared to be of the same colour as the green trees.

Dalton's work during the next six or eight years dealt chiefly with problems suggested by his meteorological observations ; he published a volume on "Meteorological Observations and Essays," chiefly occupied with descriptions of the instruments employed, more especially of the thermometer and barometer, and an instrument for determining the dew-point of air. By this time he had established the existence of a connection of some kind between magnetism and the aurora, and had thus laid the foundations of a most important branch of meteorology.

In 1799, in a note to a paper on rain and dew, he begins his work on aqueous vapour in the atmosphere by proving that water vapour exists as such in the air. This paper is quickly followed by another on the conducting power of water for heat.

A very important paper was published in 1801, on the "Constitution of Mixed Gases, etc.," wherein Dalton asserted that the total pressure of a mixture of two gases on the walls of the containing vessel is equal to the sum of the pressures of each gas; in other words, that if one gas is removed the pressure now exerted by the remaining gas is exactly the same as was exerted by that gas in the original mixture. In a paper published much later (1826), when his views and experiments on this subject were matured, he writes: "It appears to me as completely demonstrated as any physical principle, that whenever two or more . . . gases or vapours . . . are put together, either into a limited or unlimited space, they will finally be arranged each as if it occupied the whole space, and the others were not present ; the nature of the fluids and gravitation being the only efficacious agents."

This conclusion was followed out and extended in a paper published in 1803, on the absorption of gases by water and other liquids, wherein he states that the amount of each gas *mechanically dissolved* by a liquid from a mixture of gases depends only on the quantity of *that* gas in the mixture, the other gases exerting no influence in this respect.

Dalton now considered the variation in the

III.   I

pressures of various gases caused by increasing or
decreasing temperature, and then proceeded to
discuss the relations which exist between the
volumes of gases and the temperature at which
these volumes are measured. He concluded that
"all elastic fluids" under the same pressure ex-
pand equally by heat: and he adds the very important remark, "It seems, therefore, that general
laws respecting the absolute quantity and the nature
of heat are more likely to be derived from the
study of elastic fluids than of other substances"—
a remark the profound truth of which has been
emphasized by each step in the advances made in
our conception of the nature of heat since the time
of Dalton.

In these papers on the "Constitution of Mixed
Gases" Dalton also describes and illustrates a
method whereby the actual amount of water
vapour in a given bulk of atmospheric air may be
found from a knowledge of the dew-point of that
air, that is, the temperature at which the de-
position of water in the liquid form begins. The
introduction of this method for finding the humidity
of air marks an important advance in the history
of meteorology.

In this series of papers published within the
first three years of the present century Dalton evidently had before his mind's eye a picture of a gas
as a quantity of matter built up of small but
independent particles; he constantly speaks of
pressures between the small particles of elastic

fluids, of these particles as repelling each other, etc. In his "New System" he says, "A vessel full of any pure elastic fluid presents to the imagination a picture like one full of small shot."

It is very important to notice that Dalton makes use of this conception of small particles to explain purely physical experiments and operations. Although we know that during these years he was thinking much of "chemical combinations," yet we find that it was his observations on the weather which led him to the conception—a purely physical conception—of each chemically distinct gas as being built up of a vast number of small, equally heavy particles. A consideration of these papers by Dalton on the constitution of mixed gases shows us the method which he pursued in his investigations. "The progress of philosophical knowledge," he says, "is advanced by the discovery of new and important facts ; but much more when these facts lead to the establishment of general laws." Dalton always strove to attain to general laws. The facts which he describes are frequently inaccurate; he was singularly deficient in manipulation, and he cannot claim a high place as a careful experimenter. He was however able to draw general conclusions of wide applicability. He seems sometimes to have stated a generalization in definite form before he had obtained any experimental verification of it.

In the year 1802 Dalton conducted an examination of air from various localities, and concluded that one hundred volumes of air are composed of twenty-one volumes of oxygen and seventy-nine volumes of nitrogen. This appears to have been his first piece of purely chemical work. But in the next year he again returns to physical phenomena. In the paper already referred to, on the absorption of gases by water and other liquids, published in this year, he had stated that "All gases that enter into water and other liquids by means of pressure, and are wholly disengaged again by the removal of that pressure, are *mechanically* mixed with the liquid, and not *chemically* combined with it." But if this be so, why, he asked, does not water mechanically dissolve the same bulk of every kind of gas? The answer which he gives to this question is found at the close of the paper; to the student of chemistry it is very important:—

"This question I have duly considered, and though I am not yet able to satisfy myself completely, I am nearly persuaded that the circumstance depends upon the weight and number of the ultimate particles of the several gases, those whose particles are lightest and single being least absorbable, and the others more, accordingly as they increase in weight and complexity. An inquiry into the relative weights of the ultimate particles of bodies is a subject, as far as I know, entirely new. I have lately been prosecuting this inquiry with remarkable success. The principle cannot be entered upon in this paper; but I shall just subjoin the results, as far as they appear to be ascertained by my experiments." Then follows a *"Table of the relative weights of the ultimate particles of gaseous and other bodies."* The following numbers, among others, are given :—

Hydrogen 1	Sulphur ... ... 14·4
Oxygen ... ... 5·5	Alcohol ... ... 15·1
Azote ... ... 4·2	Nitrous oxide ... ... 13·7
Phosphorus ... .. 7·2	Ether .. .. 9·6

Here is the beginning of the atomic theory ; and yet Dalton's strictly chemical experimental work lies in the future. The scope of the theory is defined in that sentence—*"An inquiry into the relative weights of the ultimate particles of bodies."* His paper on mixed gases is illustrated by a plate,* which shows how vividly Dalton at this time pictured to himself a quantity of gas as composed of many little particles, and how clearly he recognized the necessity of regarding all the particles of each elementary gas as alike, but as differing from those of every other elementary gas.

In 1804 Dalton was invited to deliver a course of lectures in the Royal Institution of London, on heat, mixed gases and similar subjects. In these lectures he expounded his views on the constitution of gases, on absorption of gases by liquids, etc. These views drew much attention in this and other

* See Fig. 2, which is copied from the original in the "New System of Chemical Philosophy," and illustrates Dalton's conception of a quantity of carbonic acid gas, each atom built up of one atom of carbon and two of oxygen ; of nitrous oxide gas, each atom composed of one atom of nitrogen and one of oxygen ; and of hydrogen gas, constituted of single atoms.

countries. "They are busy with them," he writes in 1804, "at London, Edinburgh, Paris and in various parts of Germany, some maintaining one side and some another. The truth will surely out at last."

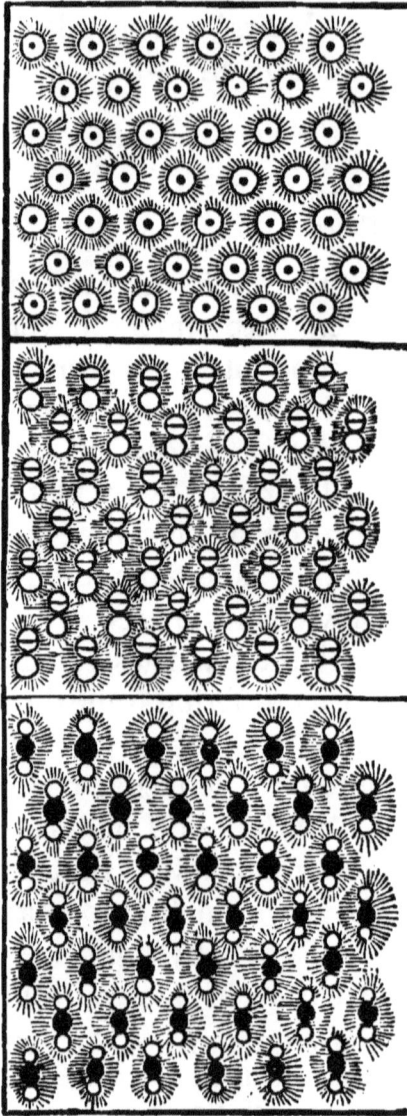

Fig. 2

Dalton's love of numerical calculations is noticeable in a trivial circumstance which he mentions in a letter from London to his brother. He tried to count the number of coaches which he met in going to the Friends' morning meeting: this he assures his brother he "effected with tolerable precision. The number was one hundred and four."

During vacation time Dalton usually made a walking excursion in the Lake district. He was extremely fond of mountain scenery, but generally combined the pursuit of science with that of pleasure; he carried his meteorological instruments with him, determined the dew-point at various altitudes, and measured mountain heights by the aid of his barometer. Sometimes however he refused to have anything to do with science. A companion in one of these excursions says that he was "like a schoolboy enjoying a holiday, mocking the cuckoos, putting up and chasing the hares, stopping from time to time to point out some beautiful view, or loitering to chat with passing pedestrians."

This side of Dalton's nature was not often apparent. In him the quiet, hard-working student generally appeared prominently marked; but on the half-holiday which he allowed himself on each Thursday afternoon, in order to enjoy the society of a few friends and to engage in his favourite amusement of a game at bowls, he laid aside something of the quietness, regularity and decorum which usually characterized him. "When it came to his turn to bowl he threw his whole soul into the game, . . . and it was not a little amusing to spectators to see him running after the ball across the green, stooping down as if talking to it, and waving his hands from one side to the other exactly as he wished the line of the ball to be, and manifesting the most intense interest in its coming near to the point at which he aimed."

From the year 1803-4 Dalton becomes more and more a worker in chemistry. The establishment of the atomic theory now engaged most of his time and attention. The results of his investigation of "the primary laws which seem to obtain in regard to heat and to chemical combinations" appeared in his "New System of Chemical Philosophy," Part I. of which, "On Heat, on the Constitution of Bodies and on Chemical Synthesis," was published in 1808.

We have now arrived at the time when Dalton's inquiry into the "relative weights of the ultimate particles of bodies" was in his opinion sufficiently advanced for presentation to the scientific world ; but I think we shall do better to postpone our consideration of this great inquiry until we have completed our review of the chief events in the life of Dalton, other than this the greatest event of all.

Dalton did not look for rewards—he desired only the just fame of one who sought for natural truths ; but after the publication of the " New System " rewards began to come to him. In 1817 he was elected a corresponding member of the French Academy of Sciences.

In 1822, when his fame as a philosophical chemist was fully established, Dalton visited Paris. This visit gave him great pleasure. He was constantly in the society of the great men who then so nobly represented the dignity of natural science in France ; Laplace, Cuvier, Biot, Arago, Gay-Lussac, Milne-Edwards and others were his friends. For some time after this visit he was more vivacious and communicative than usual, and we are told by one who lived in the same house as he, "We frequently bantered him with having become half a Frenchman." Dalton especially valued the friendship of Clementine Cuvier, daughter of the great naturalist, with whom he became acquainted during his visit to Paris. All through life he greatly delighted in the society of cultivated women, and his warmest friendships were with gentlewomen. At one time, shortly after going to Manchester, he was much taken by a widow lady who combined great personal charms with considerable mental culture. "During my *captivity*," he writes to a friend, "which lasted about a week, I lost my appetite, and had other symptoms of *bondage* about me, as incoherent discourse, etc., but have now happily regained my freedom." The society of men who like himself were actively engaged in the investigation of natural science was also a source of much pleasure to Dalton. Such men used to visit him in Manchester, so that in the house of the Rev. Mr. Johns, in whose family he lived, "there were found from time to time some of the greatest philosophers in Europe."

Dalton was elected a Fellow of the Royal Society in 1822, and four years later he became the first recipient of one of the Royal Medals, then founded by the King (George IV.). In 1830 he was elected one of the eight foreign Associates of the French Academy, an honour which is generally regarded as the highest that can be bestowed on any man of science.

Dalton was one of the original members of the British Association for the Advancement of Science, and he attended most of the meetings from the first held in York in 1831 to that held in Manchester two years before his death. At the Oxford meeting of 1832 he was created D.C.L. by the University, and two years later the University of Edinburgh honoured herself by enrolling his name on the list of her doctors of law.

About this time some of Dalton's scientific friends, who considered his work of great national importance, endeavoured to obtain a pension for him from the civil list. At the meeting of the British Association held at Cambridge in 1833, the president, Professor Sedgwick, was able to announce that "His Majesty, willing to manifest his attachment to science, and his regard for a character like that of Dr. Dalton, had graciously conferred on him, out of the funds of the civil list, a substantial mark of his royal favour." The "substantial mark of royal favour," the announcement of which Dalton received "with his customary quietness and simplicity of manner," consisted of a pension of £150 *per annum*, which was increased three years later to £300.

The second part of Volume I. of his "New System" was published by Dalton in 1810, and the second volume of the same work in 1827. In 1844 a paper by him was read before the British Association, in which he announced some important discoveries with regard to the water in crystallizable salts, and thus brought a new class of facts within the range of the atomic theory.

He was seized with paralysis in 1837, but recovered to a great extent; a second attack in 1844 however completely prostrated him. On the 16th of July in that year he made the last entry in his book of "Observations on the Weather"—"*Little rain;*" next morning he became insensible and quietly passed away.

It is as the founder of the chemical atomic theory that Dalton must ever be remembered by all students of physical and chemical science.

To the Greek philosophers Leucippus and Democritus (flourished about 440–400 B.C.) we owe the conception that ."The bodies which we see and handle, which we can set in motion or leave at rest, which we can break in pieces and destroy, are composed of smaller bodies, which we cannot see or handle, which are always in motion, and which can neither be stopped, nor broken in pieces, nor in any way destroyed or deprived of the least of their properties" (Clerk Maxwell). The heavier among these small indivisible bodies or atoms were regarded as always moving downwards. By collisions between these and the lighter ascending atoms lateral movements arose. By virtue of the natural law (as they said) that things of like weight and shape must come to the same place, the atoms of the various elements came together; thus larger masses of matter were formed; these again coalesced, and so finally worlds came into existence.

This doctrine was extended by Epicurus (340–270 B.C.), whose teaching is preserved for us in the poem of Lucretius (95–52 B.C.), "De Rerum Natura;" he ascribed to the atoms the power of deviating from a straight line in their descending motion. On this hypothesis Epicurus built a general theory to explain all material and spiritual phenomena.

The ceaseless change and decay in everything around them was doubtless one of the causes which led men to this conception of atoms as indivisible, indestructible substances which could never wear out and could never be changed. But even here rest could not be found; the mind was obliged to regard these atoms as always in motion. The dance of the dust-motes in the sunbeam was to Lucretius the result of the more complex motion whereby the atoms which compose that dust are agitated. In his dream as told by Tennyson—

> "A void was made in Nature: all her bonds
> Cracked: and I saw the flaring atom-streams
> And torrents of her myriad universe,
> Ruining along the illimitable inane,
> Fly on to clash together again, and make
> Another and another frame of things
> For ever."

The central quest of the physicist, from the days of Democritus to the present time, has been to explain the conception of "atom"—to develop more clearly the observed properties of the things which are seen and which may be handled as dependent on the properties of those things which cannot be seen, but which yet exist. For two thousand years he has been trying to penetrate beneath the ever-changing appearances of Nature, and to find some surer resting-place whence he may survey these shifting pictures as they pass before his mental vision. The older atomists thought to find this resting-place, not in the atoms themselves, but in the wide spaces which they supposed to exist between the worlds :—

> "The lucid interspace of world and world
> Where never creeps a cloud, or moves a wind,
> Nor ever falls the least white star of snow,
> Nor ever lowest roll of thunder moans,
> Nor sound of human sorrow mounts to mar
> Their sacred everlasting calm."

To the modern student of science the idea of absolute rest appears unthinkable ; but in the most recent outcome of the atomic theory—in the vortex atoms of Helmholtz and Thomson—he thinks he perceives the very " foundation stones of the material universe."

Newton conceived the atom as a "solid, massy, hard, impenetrable, movable particle." To the mind of D. Bernoulli the pressure exerted by a gas on the walls of a vessel enclosing it was due to the constant bombardment of the walls by the atoms of which the gas consisted.

Atomic motion was the leading idea in the explanation of heat given by Rumford and Davy, and now universally accepted ; and, as we have seen, Dalton was himself accustomed to regard all "elastic fluids" (*i.e.* gases) as consisting of vast numbers of atoms.

But in the year 1802 or so, Dalton thought that by the study of chemical combinations it would be possible to determine the relative weights of atoms. Assume that any elementary gas is composed of small, indivisible, equally heavy parts; assume that the weight of an atom of one element is different from that of the atom of any other element; and, lastly, assume that when elements combine the atom of the compound so produced is built up of the atoms of the various elements. Make these assumptions, and it follows that the relative weights of two or more elements which combine together must represent the relative weights of the atoms of these elements.

We know that the fixity of composition of chemical compounds had been established before this time, largely by the labours of Black and Lavoisier. Fixity of composition had however been called in question by Berthollet, who held that elements combine together in very varying quantities; that, in fact, in place of there being two or three, or a few definite compounds of, say, iron and oxygen, there exists a graduated series of such bodies; and that the amount of iron which combines with oxygen depends chiefly on such physical conditions as the temperature, the pressure, etc., under which the chemical action occurs. But by the date of the publication of the first part of Dalton's "New System," the long dispute between

Berthollet and Proust regarding fixity of composition of compounds had nearly closed in favour of the latter chemist, who strongly upheld the affirmative side of the argument.   But if Dalton's assumptions are correct, it is evident that when two elements form more than one compound, the quantity of element A in one of these must be a simple multiple of the quantity in the other of these compounds ; because there must be a greater number of atoms of element A in the atom of one compound than in that of the other compound, and an elementary atom is assumed to be indivisible.   Hence it follows that if one element be taken as a standard, it must be possible to affix to any other element a certain number which shall express the smallest quantity of that element which combines with one part by weight of the standard element ; and this number shall also represent how many times the atom of the given element is heavier than the atom of the standard element, the weight of which has been taken to be *one*.   If this element forms two compounds with the standard element, the amount of this element in the second compound must be expressed by a simple multiple of the number assigned to this element, because it is not possible, according to the fundamental assumptions of the theory, to form a compound by the combination of fractions of elementary atoms.

By pondering on the facts regarding chemical combinations which had been established by various workers previous to the year 1802, Dalton had apparently come to such conclusions as those now indicated.

In his paper on the properties of the gases constituting the atmosphere, read to the Manchester Society on November 12, 1802, he stated that one hundred measures of common air would combine with thirty-six measures of "nitrous gas" in a narrow tube to produce an oxide of nitrogen, but with seventy-two measures of the same gas in a wide vessel to produce another oxide of nitrogen. These facts, he says, "clearly point out the theory of the process : the elements of oxygen may combine with a certain portion of nitrous gas, or with twice that portion, but with no intermediate quantity."

In the concluding paragraph of his paper on absorption of gases by liquids, read on October 21, 1803, we found (see p. 116) that he had got so far in his inquiry into the "relative weights of the ultimate particles of bodies" as to give a table of twenty-one such weights. About this time Dalton made analyses of two gaseous compounds of carbon—olefiant gas and carburetted hydrogen or marsh-gas. He found that both are compounds of carbon and hydrogen ; that in one 4·3 parts by weight of carbon are combined with one part by weight of hydrogen, and in the other the same amount (4·3) of carbon is combined with two parts by weight of hydrogen.*

---

* More accurate analysis has shown that there are six parts of carbon united respectively with one and with two parts by weight of hydrogen in these compounds.

This was a striking confirmation of his views regarding combination in multiple proportions, which views followed as a necessary deduction from the atomic hypothesis. From this time he continued to develop and extend this hypothesis, and in the year 1808 he published his "New System of Chemical Philosophy."

The first detailed account of the atomic theory was however given to the chemical world the year before Dalton's book appeared. During a conversation with Dalton in the autumn of 1804 Dr. Thomas Thomson learned the fundamental points of the new theory, and in the third edition of his "System of Chemistry," published in 1807, he gave an account of Dalton's views regarding the composition of bodies.

In the same year a paper by Thomson appeared in the *Philosophical Transactions*, wherein it was experimentally proved that oxalic acid combines with strontia to form two distinct compounds, one of which contains twice as much oxalic acid as the other, the amount of strontia being the same in both. Analyses of the oxalates of potash, published about the same time by Wollaston, afforded another illustration of the *law of multiple proportions*, and drew the attention of chemists to Dalton's theory. But the new theory was opposed by several very eminent chemists, notably by Sir Humphry Davy. In the autumn of 1807 Wollaston, Thomson and Davy were present at the dinner of the Royal Society Club, at the Crown and

III.   K

Anchor, in the Strand. After dinner, these three chemists' discussed the new theory for an hour and a half, Wollaston and Thomson trying to convince Davy of the truth of Dalton's theory; but " so far from being convinced, he went away, if possible, more prejudiced against it than ever."

Soon after this Wollaston succeeded in convincing Mr. Davis Gilbert (afterwards President of the Royal Society) of the justness of the atomic theory, and he in turn so placed the facts and the reasoning before Davy, that from this time he became a supporter of the new theory.

In order that the atomic theory should be fruitful of results, it was now necessary that the values of the atomic weights of many elements should be carefully determined.

Let us consider what knowledge must be acquired before the value to be assigned to the atomic weight of an element can be found.

Hydrogen was the element chosen as a standard by Dalton. He assumed that the atom of hydrogen weighs 1 ; the atomic weight of any other element is therefore a number which tells how many times the atom of that element is heavier than the atom of hydrogen. Thus, when Dalton said the atomic weight of oxygen is 8, he meant that the atom of oxygen is eight times heavier than that of hydrogen. How was this number obtained?

Accurate analyses of water show that in this liquid one part by weight of hydrogen is combined with eight parts by weight of oxygen; but (it is

said) as the atom of hydrogen weighs 1, the atom
of oxygen must weigh 8. In drawing this conclusion it is assumed that the atom, or smallest
particle, of water is built up of one atom of hydrogen
and one atom of oxygen. Let it be assumed that
the atom of water contains two atoms of hydrogen
and one of oxygen, then the latter atom must weigh
sixteen times as much as each atom of hydrogen;
let it be assumed that three atoms of hydrogen
combine with one atom of oxygen to form an
atom of water, then the weight of the oxygen atom
must be twenty-four times that of the hydrogen
atom. Any one of these assumptions will equally
satisfy the figures obtained by analyzing water
($1 : 8 = 2 : 16 = 3 : 24$). Now, had we any
method whereby we could determine how many
times an atom of water is heavier than an atom of
hydrogen we should be able to determine which
of the foregoing assumptions is correct, and therefore to determine the atomic weight of oxygen.
Hence, before the atomic weight of an element can
be determined, there must be found some method
for determining the atomic weights of compounds
of that element. Unless this can be done the
atomic theory is of little avail in chemistry.

I conceive it to be one of the signal merits of
Dalton that he so clearly lays down rules, the best
which could be devised at his time, for determining the atomic weights of compounds, or, what is
the same thing, for determining the number of
elementary atoms in one atom of any compound.

In his " New System " he says that he wishes to show the importance of ascertaining " the relative weights of the ultimate particles both of simple and compound bodies, the number of simple elementary particles which constitute one compound particle, and the number of less compound particles which enter into the formation of one more compound particle."

Considering compounds of two elements, he divides these into binary, ternary, quaternary, etc., according as the compound atom contains two, three, four, etc., atoms of the elements. He then proceeds thus—

" The following general rules may be adopted as guides in all our investigations respecting chemical synthesis :—

" 1st. When only one combination of two bodies can be obtained, it must be presumed to be a *binary* one, unless some cause appear to the contrary.

" 2nd. When two combinations are observed, they must be presumed to be a *binary* and a *ternary*.

" 3rd. When three combinations are obtained, we may expect one to be *binary* and the other two *ternary*.

" 4th. When four combinations are observed, we should expect one *binary*, two *ternary*, and one *quaternary*," etc.

Only one compound of hydrogen and oxygen was then known ; hence it was presumed to be a binary compound, *i.e.* a compound the smallest particle of which consisted of one atom of hydrogen and one atom of oxygen ; and hence, from the data already given on page 130, it followed that the atomic weight of oxygen was 8.   Two compounds of carbon and oxygen were known, each containing six parts by weight of carbon, in one case united with eight, and in the other case with sixteen parts by weight of oxygen.   From Dalton's rules one of these was a binary, and the other a ternary compound ; but as the atomic weight of oxygen had already been determined to be 8, that compound of carbon and oxygen containing eight of oxygen combined with six of carbon was decided to be binary, and that containing sixteen of oxygen (*i.e.* two atoms) to be ternary ; and hence the atomic weight of carbon was determined to be 6.

In the second part of the " New System " Dalton, guided by these rules, determined experimentally the atomic weights of a great many substances ; but this was not the kind of work suited to Dalton's genius.   His analytical determinations were generally inaccurate; nevertheless, he clearly showed how the values of the atomic weights of elements ought to be established, and he obtained results sufficiently accurate to confirm his general theory. To make accurate determinations of the relative weights of elementary atoms was one of the tasks reserved for the great Swedish chemist Berzelius (see pp. 162–170).   When we examine Dalton's rules we must confess that they appear somewhat arbitrary. He does not give reasons for his assertion that "when only one combination of two bodies can be obtained, it must be presumed to be a binary one." Why may it not be ternary or quaternary? Why must the atom of water be built up of one atom of hydrogen combined with one atom of oxygen? Or, when two compounds are known containing the same pair of elements, why must one be binary and the other ternary?

Or, even assuming that this *must* be justified by facts, does it follow that Dalton's interpretation of the atomic structure of the two oxides of carbon is necessarily correct? These oxides contain 6 of carbon + 8 of oxygen, and 6 of carbon + 16 of oxygen, respectively.

Take the second, $6 : 16 = 3 : 8$; assume this to be a binary compound of one atom of oxygen (weighing 8) with one atom of carbon (weighing 3), then the other will be a ternary compound containing one atom of oxygen (8) and two atoms of carbon (6).

Hence it appears that Dalton's rules were too arbitrary, and that they were insufficient to determine with certainty the atomic weights of some of the elements. Nevertheless, without some such rules as those of Dalton, no great advances could have been made in applying the atomic theory to the facts of chemical combination; and Dalton's rules were undoubtedly founded on wide considerations. In the appendix to Volume II. of his "New System" he expressly states that before the number of atoms of two elements present in the

atom of a compound can be determined, it is necessary that many combinations should be examined,
not only of these elements with each other, but also
of each of these with other elements ; and he tells us
that to gather together facts bearing on this general
question of chemical synthesis was the object of
his work from the time of the promulgation of the
atomic theory.

When we find that Dalton applied the term
" atom " to the small particles of compound bodies,
we at once see that by atom he could not always
mean " that which cannot be cut ; " he simply
meant the smallest particle of a substance which
exhibits the properties of that substance.

A mass of water vapour was conceived by Dalton
as " like a mass of small shot."   Each shot exhibited
the characteristic chemical properties of water
vapour ; it differed from the large quantity of vapour
only in mass ; but if one of these little pieces of
shot were divided—as Dalton, of course, knew it
could be divided—smaller pieces of matter would
be produced.   But these would no longer be water ;
they would be new kinds of matter.   They are called
oxygen and hydrogen.

As aids towards gaining a clear conception of
the " atom " of a compound as a definite building,
Dalton made diagrammatic representations of the
hypothetical structures of some of these atoms :
the following plate is copied from the " New
System : "—A represents an atom of alum ; B, an
atom of nitrate of alumina ; C, of barium chloride ;

D, of barium nitrate ; E, of calcium chloride ; F, of calcium nitrate ; G, of calcium sulphate ; II, of

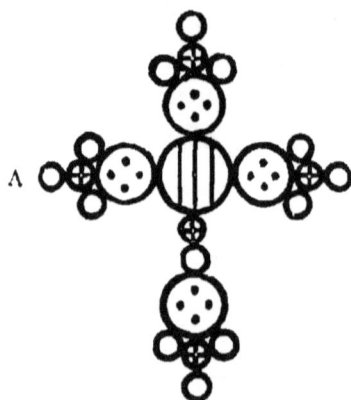

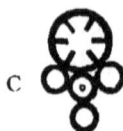

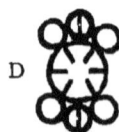

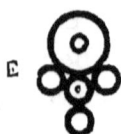

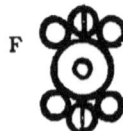

Fig. 3.

potassium carbonate ; I, of potash ; and K, an atom of soda.

But I think if we consider this application of the term " atom " to elements and compounds alike, we shall see objections to it. When an atom of a compound is divided the smaller particles so produced are each very different in chemical properties from the atom which has just been divided. We may, if we choose, assume that the atom of an element could in like manner be divided, and that the products of this division would be different from the elementary atoms; but such a division of an elementary atom has not as a matter of fact been yet accomplished, unless we class among elements substances such as potash and soda, which for many years were universally regarded as elements, and rightly so regarded because they had not been decomposed. In Dalton's nomenclature then, the term " atom " is applied alike to a small particle with definite properties known to be divisible into smaller particles, each with properties different from those of the undivided particle, and to a small particle which, so far as our knowledge goes, cannot be divided into any particle smaller than or different from itself.

Nevertheless, if the atomic theory was to be victorious, it was necessary that it should be applied to elements and compounds alike. Until a clear conception should be obtained, and expressed in accurate language, of the differences in structure of the ultimate particles of compounds and of elements, it was perhaps better to apply the term "atom" to both alike.

These two difficulties—(1) the difficulty of attaching to the term "atom" a precise meaning applicable to elements and compounds alike, and (2) the difficulty of determining the number of elementary atoms in the atom of a given compound, and hence of determining the relative weights of elementary atoms themselves—were for many years stumbling-blocks in the path of the upholders of the Daltonian theory.

The very great difficulty of clearly comprehending the full meaning of Dalton's proposed theory becomes apparent when we learn that within three years from the publication of Part I. of the "New System," facts were made known by the French chemist Gay-Lussac, and the true interpretation of these facts was announced by the Italian chemist Avogadro, which facts and interpretation were sufficient to clear away both the difficulties I have just mentioned; but that nevertheless it is only within the last ten or fifteen years that the true meaning of the facts established by Gay-Lussac and the interpretation given by Avogadro have been generally recognized.

In 1809 Gay-Lussac, in a memoir on the combination of gaseous bodies, proved that gases combine chemically in simple proportions by volume, and that the volume of the product always bears a simple relation to the volumes of the combining gases. Thus, he showed that two volumes of hydrogen combine with one volume of oxygen to form two volumes of water vapour ; that one volume of nitrogen combines with three volumes of hydrogen to form two volumes of ammonia gas, and so on. Now, as elements combine atom with atom, the weights of these combining volumes of elements must represent the relative weights of the atoms of the same elements.

In 1811 Avogadro distinguished between the ultimate particles of compounds and elements. Let a gaseous element, A, combine with another gaseous element, B, to form a gaseous compound, C ; then Avogadro supposed that the little particles of A and the little particles of B (Dalton's atoms) split up, each into two or more smaller particles, and that these smaller particles then combine together to form particles of the compound C. The smaller particles produced by splitting a Daltonian elementary atom were regarded by Avogadro as all identical in properties, but these very small particles could not exist uncombined either with each other or with very small particles of some other element. When the atom of a compound is decomposed, Avogadro pictured this atom as splitting into smaller particles of two or three or more different kinds, according as the compound had contained two or three or more different elements.

To Avogadro's mental vision an elementary gas appeared as built up of a great many little particles, each exhibiting in miniature all the properties of the gas. The gas might be heated, or cooled, or otherwise physically altered, but each of the little particles remained intact ; the moment however that this gas was mixed with another on which it could chemically react, these little particles split into smaller parts, but as the smaller parts so produced could not exist in this state, they seized hold of the corresponding very small parts of the other gas, and thus a particle of a compound gas was produced.

A compound gas was pictured by Avogadro as also built up of small particles, each exhibiting in miniature the properties of the gas, and each remaining undecomposed when the gas was subjected only to physical actions ; but when the gas was chemically decomposed, each little particle split, but the very small parts thus produced, being each a particle of an elementary substance, continued to exist, and could be recognized by the known properties of that element.

To the smallest particle of any substance (elementary or compound) which exhibits the properties of that substance, and which cannot be split into parts without destroying these properties, we now give the name of *molecule*.

A molecule is itself a structure. It is built up of parts ; each of these parts we now call an *atom*. The molecule of a compound is, of course, composed of the atoms of the elements which form that compound. The molecule may contain two or three or more unlike atoms. The molecule of an element is composed of the atoms of that element, and all of these atoms are supposed to be alike. We cannot get hold of elementary atoms and examine them, but we have a large mass of evidence in favour of the view which regards the molecule of an element as composed of parts each weighing less than the molecule itself.

The student of physics or chemistry now believes that, were a very small quantity of a gas (say ammonia) or a drop of a liquid (say water) magnified to something like the size of the earth, he should see before him a vast heap of particles of ammonia or of water, each exhibiting all the properties by the possession of which he now distinguishes ammonia or water from all other kinds of matter. He believes that he should see these particles in motion, each moving rapidly from place to place, sometimes knocking against another, sometimes traversing a considerable space without coming into collision with any other. But the student tries to penetrate yet further into the nature of things. To the vision of the chemist these particles of almost inconceivable minuteness are themselves built up of smaller particles. As there is an architecture of masses, so is there an architecture of molecules. Hydrogen and oxygen are mixed ; the chemist sees the molecules of each in their never-ceasing dance moving here and there among the molecules of the other, yet each molecule retaining its identity ; an electric spark is passed through the mixture, and almost instantaneously he sees each hydrogen molecule split into two parts, and each oxygen molecule split into two parts, and then he sees these parts of molecules, these atoms, combine, a pair of hydrogen atoms with an atom of oxygen, to form compound molecules of water.

Avogadro's hypothesis gave the chemist a definition of " molecule ; " it also gave him a definition of " atom."

It is evident that, however many atoms of a given element there may be in this or in that compound molecule, no compound of this element can exist containing less than a single atom of the element in question ; therefore an atom of an element is the smallest quantity of that element in the molecule of any compound thereof.

And so we have come back to the original hypothesis of Dalton ; but we have extended and modified that hypothesis—we have distinguished two orders of small particles, the molecule (of a compound or of an element) and the atom (of an element). The combination of two or more elements is now regarded as being preceded by the decomposition of the molecules of these elements into atoms. We have defined molecule and we have defined atom, but before we can determine the relative weights of elementary atoms we must have a means of determining the relative weights of compound molecules. The old difficulty still stares us in the face—how can we find the number of elementary atoms in the molecule of a given compound ?

The same naturalist who enriched chemical science by the discovery of the molecule as distinct from the atom, placed in the hands of chemists the instrument for determining the relative weights of molecules, and thus also the relative weights of atoms.

The great generalization, usually known as *Avogadro's law*, runs thus : "*Equal volumes of gases measured at the same temperature and under the same pressure contain equal numbers of molecules.*"

Gay-Lussac had concluded that " equal volumes of gases contain equal numbers of atoms ; " but this conclusion was rejected, and rightly rejected by Dalton, who however at the same time refused to admit that there is a simple relation between the combining volumes of elements. The generalization of Avogadro has however stood the test of experiment, and is now accepted as one of the fundamental " laws " of chemical science.

Like the atomic theory itself, Avogadro's law is an outcome of physical work and of physical reasoning. Of late years the great naturalists, Clausius, Helmholtz, Joule, Rankine, Clerk Maxwell and Thomson have developed the physical theory of molecules, and have shown that Avogadro's law may be deduced as a necessary consequence from a few simple physical assumptions. This law has thus been raised, from being a purely empirical generalization, to the rank of a deduction from a wide, yet simple physical theory.

Now, if " equal volumes of gases contain equal

numbers of molecules," it follows that the ratio of
the densities of any two gases must also be the
ratio of the weights of the molecules which constitute these gases. Thus, a given volume of water
vapour weighs nine times more than an equal
volume of hydrogen ; therefore the molecule of
gaseous water is nine times heavier than the molecule of hydrogen. One has therefore only to
adopt a standard of reference for molecular
weights, and Avogadro's law gives the means
of determining the number of times any gaseous
molecule is heavier than that of the standard
molecule.

But consider the combination of a gaseous
element with hydrogen ; let us take the case of
hydrogen and chlorine, which unite to form gaseous
hydrochloric acid, and let us determine the volumes
of the uniting elements and the volume of the
product. Here is a statement of the results : one
volume of hydrogen combines with one volume of
chlorine to form two volumes of hydrochloric acid.
Assume any number of molecules we please in the
one volume of hydrogen—say ten—there must be,
by Avogadro's law, also n molecules in the one
volume of chlorine ; but inasmuch as the volume of
hydrochloric acid prod is double that of either
the hydrogen or the chlorine which combined to
form it, it follows, by the same law, that twenty
molecules of hydrochloric acid have been formed
by the union of ten molecules of hydrogen with
ten molecules of chlorine. The necessary conclu-

sion is that each hydrogen molecule and each chlorine molecule has split into two parts, and that each half-molecule (or atom) of hydrogen has combined with one half-molecule (or atom) of chlorine, to produce one compound molecule of hydrochloric acid.

Therefore we conclude that the hydrogen molecule is composed of two atoms, and that the chlorine molecule is also composed of two atoms; and as hydrogen is to be our standard element, we say that if the atom of hydrogen weighs one, the molecule of the same element weighs two.

It is now easy to find the *molecular weight* of any gas; it is only necessary to find how many times heavier the given gas is than hydrogen, the weight of the latter being taken as 2. Thus, oxygen is sixteen times heavier than hydrogen, but $1 : 16 = 2 : 32$, therefore the molecule of oxygen is thirty-two times heavier than the molecule of hydrogen. Ammonia is eight and a half times heavier than hydrogen, but $1 : 8\frac{1}{2} = 2 : 17$, therefore the molecule of ammonia is seventeen times heavier than the molecule of hydrogen. This is what we more concisely express by saying " the molecular weight of oxygen is 32," or "the molecular weight of ammonia is 17," etc., etc.

Now, we wish to determine the *atomic weight* of oxygen; that is, we wish to find how many times the oxygen atom is heavier than the atom of hydrogen. We make use of Avogadro's law and of the definition of " atom " which has been deduced from it (see p. 142).

III.   I.

We know that eight parts by weight of oxygen combine with one part by weight of hydrogen to form water; but we do not know whether the molecule of water contains one atom of each element, or two atoms of hydrogen and one atom of oxygen, or some other combination of these atoms (see p. 131). But by vaporizing water and weighing the gas so produced, we find that water vapour is nine times heavier than hydrogen: now, $1 : 9 = 2 : 18$, therefore the molecular weight of water gas is 18. Analysis tells us that eighteen parts by weight of water gas contain sixteen parts of oxygen and two parts of hydrogen; that is to say, we now know that in the molecule of water gas there are two atoms of hydrogen combined with sixteen parts by weight of oxygen. We now proceed to analyze and determine the molecular weights of as many gaseous compounds of oxygen as we can obtain. The outcome of all is that we have as yet failed to obtain any such compound in the molecule of which there are less than sixteen parts by weight of oxygen. In some of these molecules there are sixteen, in some thirty-two, in some forty-eight, in some sixty-four parts by weight of oxygen, but in none is there less than sixteen parts by weight of this element. Therefore we conclude that the atomic weight of oxygen is 16, because this is the smallest amount, referred to hydrogen taken as 1, which has hitherto been found in the molecule of any compound of oxygen.

The whole of the work done since the publication of Dalton's "New System" has emphasized the importance of that chemist's remark, that no safe conclusion can be drawn as to the value of the atomic weight of an element except from a consideration of many compounds of that with other elements.   But in Avogadro's law we have a far more accurate and trustworthy method for determining the molecular weights of compounds than any which Dalton was able to devise by his study of chemical combinations.

We have thus got a clearer conception of "atom" than was generally possessed by chemists in the days of Dalton, and this we have gained by introducing the further conception of "molecule" as that of a quantity of matter different from, and yet similar to, the atom.

The task now before us will for the most part consist in tracing the further development of the fundamental conception of Dalton, the conception, viz., of each chemical substance as built up of small parts possessing all the properties, other than the mass, of the whole ; and—what we also owe to Dalton—the application of this conception to explain the facts of chemical combination.

The circumstances of Dalton's early life obliged him to trust largely to his own efforts for acquiring knowledge ; and his determination not to accept facts at second hand but to acquire them for himself, is very marked throughout the whole of his life.   In the preface to the second part of the "New System" he says, "Having been in my progress so often misled by taking for granted the results of others, I have determined to write as little as possible but what I can attest by my own experience."

We should not expect such a man as this to make any great use of books ; one of his friends tells us that he heard him declare on a public occasion that he could carry his library on his back, and yet had not read half of the books which comprised it.

The love of investigation which characterized Dalton when young would naturally be increased by this course of intellectual life. How strong this desire to examine everything for himself became, is amusingly illustrated by a story told by his medical adviser, Dr. Ransome. Once when Dalton was suffering from catarrh Dr. Ransome had pre-scribed a James's powder, and finding his patient much better next day, he congratulated himself and Dalton on the good effects of the medicine. " I do not well see how that can be," said Dalton, " as I kept the powder until I could have an oppor-tunity of analyzing it."

As Dalton grew older he became more than ever disinclined to place much trust in the results obtained by other naturalists, even when these men were acknowledged to be superior to himself in manipulative and experimental skill. Thus, as we have already learned, he could not be brought to allow the truth of Gay-Lussac's experimentally established law regarding gaseous combinations ; he preferred to attribute Gay-Lussac's results to errors of experiment. "The truth is, I believe, that gases do not unite in equal or exact measures in any one instance; when they appear to do so it is owing to the inaccuracy of our experiments."

That Dalton did not rank high as an experimenter is evident from the many mistakes in matters of fact which are to be found in the second part of his "New System." A marked example of his inaccuracy in purely experimental work is to be found in the supposed proof given by him that charcoal, after being heated to redness, does not absorb gases. He strongly heated a quantity of charcoal, pulverized it, and placed it in a Florence flask, which was connected by means of a stopcock with a bladder filled with carbonic acid: after a week he found that the flask and its contents had not sensibly increased in weight, and he concluded that no carbonic acid had been absorbed by the charcoal. But no trustworthy result could be obtained from an experiment in which the charcoal, having been deprived of air by heating, was again allowed to absorb air by being pulverized in an open vessel, and was then placed in a flask filled with air, communication between the carbonic acid and the external air being prevented merely by a piece of bladder, a material which is easily permeated by gases.

Dalton used a method which can only lead to notable results in natural science when employed by a really great thinker; he acquired a few facts, and then thought out the meaning of these.

Almost at the beginning of each investigation he tried to get hold of some definite generalization, and *then* he proceeded to amass special facts. The object which he kept before himself in his experimental work was to establish or to disprove this or that hypothesis. Every experiment was conducted with a clearly conceived aim. He was even willing to allow a large margin for errors of experiment if he could thereby bring the results within the scope of his hypothesis.

That the *law of multiple proportions* is simply a generalization of facts, and may be stated apart from the atomic theory, is now generally admitted. But in Dalton's mind this law seems to have arisen rather as a deduction from the theory of atoms than to have been gained as a generalization from experiments. He certainly always stated this law in the language of the atomic theory. In one of his walking excursions he explained his theory to a friend, and after expounding his views regarding atomic combinations, he said that the examples which he had given showed the necessary existence of the principle of multiple proportions: "Thou knowest it must be so, for no man can split an atom." We have seen that carburetted hydrogen was one of the compounds on the results of the analysis of which he built his atomic theory; yet we find him saying of the constitution of this compound that "no correct notion seems to have been formed till the atomic theory was introduced and applied in the investigation."

When Dalton was meditating on the laws of chemical combination, a French chemist, M. Proust, published analyses of metallic oxides, which proved that when a metal forms two oxides the amount of metal in each is a fixed quantity—that there is a sudden jump, as it were, from one oxide to another. We are sometimes told that from these experiments Proust would have recognized the law of multiple proportions had his analyses only been more accurate; but we know that Dalton's analyses were very inaccurate, and yet he not only recognized the law of multiple proportions, but propounded and established the atomic theory. Something more than a correct system of keeping books and balancing accounts is wanted in natural science. Dalton's experimental results would be the despair of a systematic analyst, but from these Dalton's genius evolved that splendid theory which has done so much to advance the exact investigation of natural phenomena.

Probably no greater contrast could be found between methods of work, both leading to the establishment of scientific (that is, accurate and precise) results, than that which exists between the method of Dalton and the method pursued by Priestley.

Priestley commenced his experiments with no particular aim in view; sometimes he wanted to amuse himself, sometimes he thought he might light upon a discovery of importance, sometimes his curiosity incited him to experiment. When he

got facts he made no profound generalizations;
he was content to interpret his results by the help
of the prevailing theory of his time. But each new
fact only spurred him on to make fresh incursions
into the fields of Nature. Dalton thought much and
deeply; his experimentally established facts were
to him symbols of unseen powers. He used facts
as Hobbes says the wise man uses words: they
were his counters only, not his money.

When we ask how it was that Dalton acquired his
great power of penetrating beneath the surface
of things and finding general laws, we must
attribute this power in part to the training which
he gave himself in physical science. It was from a
consideration of physical facts that he gained the
conception of ultimate particles of definite weight.
His method was essentially dynamical; that is, he
pictured a gas as a mass of little particles, each
of which acted on and was acted on by, other
particles. The particles were not thrown together
anyhow; definite forces existed between them.
Each elementary or compound gas was pictured
as a system of little particles, and the properties
of that gas were regarded as dependent on the
nature and arrangement of these particles. Such
a conception as this could only be gained by a
careful and profound thinker versed in the methods
of physical and mathematical science. Thus we
see that although Dalton appeared to gain his
great chemical results by a method which we are
not generally inclined to regard as the method

of natural science, yet it was by virtue of his careful training in a branch of knowledge which deals with facts, as well as in that science which deduces particular conclusions from general principles, that he was able to introduce his fruitful conceptions into the science of chemistry.

To me it appears that Dalton was pre-eminently distinguished by the possession of imagination. He formed clear mental images of the phenomena which he studied, and these images he was able to combine and modify so that there resulted a new image containing in itself all the essential parts of each separate picture which he had previously formed.

From his intense devotion to the pursuit of science the development of Dalton's general character appears to have been somewhat dwarfed. Although he possessed imagination, it was the imagination of a naturalist rather than that of a man of broad culture. Perhaps it was a want of broad sympathies which made him trust so implicitly in his own work and so readily distrust the work of others, and which moreover led him astray in so many of his purely experimental investigations.

Dalton began his chemical work about six years after the death of Lavoisier. Unlike that great philosopher he cared nothing for political life. The friends in whose family he spent the greater part of his life in Manchester were never able to tell whether he was Whig or Tory. Unlike Priestley he was content to let metaphysical and theological speculation alone. In his quiet devotion to study he more resembled Black, and in his method, which was more deductive than that usually employed in chemistry, he also resembled the Edinburgh professor. Trained from his earliest days to depend on himself, nurtured in the creedless creed of the Friends, he entered on his life's work with few prejudices, if without much profound knowledge of what had been done before him. By the power of his insight into Nature and the concentration of his thought, he drew aside the curtain which hung between the seen and the unseen ; and while Herschel, sweeping the heavens with his telescope and night by night bringing new worlds within the sphere of knowledge, was overpowering men's minds by new conceptions of the infinitely great, John Dalton, with like imaginative power, was examining the architecture of the ultimate particles of matter, and revealing the existence of law and order in the domain of the infinitely small.

CHAPTER IV.

ESTABLISHMENT OF GENERAL PRINCIPLES OF CHEMICAL SCIENCE (*continued*)—PERIOD OF DAVY AND BERZELIUS.

Humphry Davy, 1778–1829. *Johann Jacob Berzelius*, 1779–1848.

WE may roughly date the period of chemical advance during which the connections between chemistry and other branches of natural knowledge were recognized and studied, as beginning with the first year of this century, and as continuing to our own day.

The elaboration of the atomic theory was busily carried on during the second and third decades of this century ; to this the labour of the Swedish chemist Berzelius largely contributed.

That there exist many points of close connection between chemical and electrical science was also demonstrated by the labours of the same chemist, and by the brilliant and impressive discoveries of Sir Humphry Davy.

A system of classification of chemical elements and compounds was established by the same great naturalists, and many inroads were made into the domain of the chemistry of bodies of animal and vegetable origin.

The work of Berzelius and Davy, characterized as it is by thoroughness, clearness and definiteness, belongs essentially to the modern era of chemical advance; but I think we shall better preserve the continuity of our story if we devote a chapter to a consideration of the work of these two renowned naturalists before entering on our review of the time immediately preceding the present, as typical workers in which time I have chosen Liebig and Dumas.

In the last chapter we found that the foundations of the atomic theory had been laid, and the theory itself had been applied to general problems of chemical synthesis, by Dalton. In giving, in that chapter, a short sketch of the modern molecular theory, and in trying to explain the meaning of the term " molecule " as contrasted with "atom," I necessarily carried the reader forward to a time considerably later than the first decade of this century. We must now retrace our steps; and in perusing the account of the work of Berzelius and Davy given in the present chapter, the reader must endeavour to have in his mind a conception of atom analogous to the mental picture formed by Dalton (see pp. 135, 136); he must regard the term as applicable to element and compound alike;

he must remember that the work of which he reads is the work of those who are striving towards a clear conception of the atom, and who are gradually rising to a recognition of the existence of more than one order of small particles, by the regular putting together of which masses of matter are constituted.

No materials, so far as I am aware, exist from which a life of Berzelius can be constructed. I must therefore content myself with giving a mere enumeration of the more salient points in his life. Of his chemical work abundant details are fortunately to be found in his own "Lehrbuch," and in the works and papers of himself and his contemporaries.

JOHANN JACOB BERZELIUS was the son of the schoolmaster of Wäfersunda, a village near Linköping, in East Gothland, Sweden. He was born in August 1779—he was born, that is, a few years after Priestley's discovery of oxygen ; at the time when Lavoisier had nearly completed his theory of combustion ; when Dalton was endeavouring to keep the unruly youth of Eaglesfield in subjection ; and when Black, having established the existence of fixed air and the theory of latent heat, was the central figure in the band of students who were enlarging our knowledge of Nature in the Scottish capital.

Being left an orphan at the age of nine, the young Berzelius was brought up by his grandfather, who appears to have been a man of education and sense. After attending school at Linköping, he entered the University of Upsala as a student of medicine. Here he soon began to show a taste for chemistry. It would appear that few or no experiments were then introduced into his lectures by the Professor of Chemistry at Upsala; little encouragement was given to pursue chemical experiments, and so Berzelius had to trust to his own labours for gaining an acquaintance with practical chemistry. Having thus made considerable progress in chemistry, and being on a visit to the mineral baths of Medevi, he seized the opportunity to make a very thorough analysis of the waters of this place, which were renowned in Sweden for their curative properties. The publication of this analysis marks the first appearance of Berzelius as an author.

He graduated as M.B. in 1801, and a year or two later presented his dissertation, entitled "The Action of Galvanism on Organic Bodies," as a thesis for the degree of Doctor of Medicine. This thesis, like that of Black, published about half a century earlier, marks an important stage in the history of chemistry. These and other publications made the young doctor famous; he was called to Stockholm to be extraordinary (or assistant) Professor of Chemistry in the medical school of that capital.

Sometimes practising medicine in order to add to his limited income, but for the most part engaged in chemical research, he remained in Stockholm for nearly fifty years, during most of which time the laboratory of Berzelius in the Swedish capital was regarded as one of the magnetic poles of the chemical world. To this point came many of the great chemists who afterwards enriched the science by their discoveries. Wöhler, H. and G. Rose, Magnus, Gmelin, Mitscherlich and others all studied with Berzelius. He visited England and France, and was on terms of intimacy and in correspondence with Davy, Dalton, Gay-Lussac, Berthollet and the other men who at that period shed so much lustre on English and French science.

It is said that Berzelius was so much pleased with the lectures of Dr. Marcet at Guy's Hospital, that on his return from his visit to England in 1812, he introduced much more liveliness and many more experimental illustrations into his own lectures.

At the age of thirty-one, Berzelius was chosen President of the Stockholm Academy of Sciences ; a few years later he was elected a Foreign Fellow of the Royal Society, which society bestowed on him the Copley Medal in 1836. He was raised to the rank of a baron by the King of Sweden, being allowed as a special privilege to retain his own name.

In the year 1832 Berzelius resigned his professorship, and in the same year he married. During the remainder of his life, he continued to receive honours of all kinds, but he never for a moment forsook the paths of science. After the death of Davy, in 1829, he was recognized as the leading

European chemist of his age; but, although firm in his own theoretical views, he was ready to test these views by appealing to Nature. The very persistency with which he clung to a conception established on some solid experimental basis insured that new light would be thrown on that conception by the researches of those chemists who opposed him.

Probably no chemist has added to the science so many carefully determined facts as Berzelius; he was always at work in the laboratory, and always worked with the greatest care. Yet the appliances at his command were what we should now call poor, meagre, and utterly inadequate. Professor Wöhler of Göttingen, who in the fulness of days and honours has so lately gone from amongst us, recently gave an account of his visit to Berzelius in the year 1823. Wöhler had taken his degree as Doctor of Medicine at Heidelberg, and being anxious to prosecute the study of chemistry he was advised by his friends to spend a winter in the laboratory of the Swedish professor. Having written to Berzelius and learned that he was willing to allow him working room in his laboratory, the young student set out for Stockholm. After a journey to Lübeck and a few days' passage in a small sailing-vessel, he arrived in the Swedish capital.

Knocking at the door of the house pointed out as that of Berzelius, he tells us that his heart beat hard as the door was opened by a tall man of florid complexion. "It was Berzelius himself," he exclaims. Scarcely believing that he was in the very room where so many famous discoveries had been made, he entéred the laboratory. No water, no gas, no draught-places, no ovens were to be seen; a couple of plain tables, a blowpipe, a few shelves with bottles, a little simple apparatus, and a large water-barrel whereat Anna, the ancient cook of the establishment, washed the laboratory dishes, completed the furnishings of this room, famous throughout Europe for the work which had been done in it. In the kitchen which adjoined, and where Anna cooked, was a small furnace and a sand bath for heating purposes.

In this room many great discoveries were made. Among these we may note the separation of the element columbium in 1815, and of selenion in 1818; the discovery of the new earth thoria in 1828; the elucidation of the properties of yttrium and cerium about 1820, of uranium in 1823, and of the platinum metals in 1828; the accurate determination of the atomic weights of the greater number of the elements; the discovery of "sulphur salts" in 1826-27, and the proof that silica is an acid, and that most of the "stony" minerals are compounds of this acid with various bases.

But we shall better learn the value of some of these discoveries by taking a general review of the contributions to chemical science of the man who spent most of his life at work in that room in Stockholm.

III.   M

The German chemist Richter, in the first or second year of this century, had drawn attention to the fact that when two neutral compounds, such as nitrate of potash and chloride of lime, react chemically, the substances produced by this reaction are also neutral. All the potash combined with nitric acid in one salt changes places with all the lime combined with muriatic acid in the other salt; therefore, said Richter, these different quantities of potash and lime are neutralized by the same quantity of nitric acid; and, hence, these amounts of potash and lime are chemically *equivalent*, because these are the amounts which perform the same reaction, viz. neutralization of a fixed quantity of acid. If then careful analyses were made of a number of such neutral compounds as those named, the *equivalents* of all the commoner " bases " and " acids " * might be calculated.

Richter's own determinations of the equivalents of acids and bases were not very accurate, but Berzelius was impressed with the importance of this work. The year before the appearance of Dalton's " New System " (*i.e.* in 1807), he began to prepare and carefully analyze series of neutral salts. As the work was proceeding he became acquainted with the theory of Dalton, and at once saw its extreme importance. For some time Berzelius continued to work on the lines laid down by Dalton, and to accumulate data from which the

---

* The history and meaning of these terms is considered on p. 171, *et seq.*

atomic weights of elements might be calculated ;
but he soon perceived—as the founder of the theory
had perceived from the very outset—that the fundamental conception of each atom of an element
as being a distinct mass of matter weighing more
or less than the atom of every other element, and
of each atom of a compound as being built up of
the atoms of the elements which compose that
compound,—Berzelius, I say, perceived that these
conceptions must remain fruitless unless means
were found for determining the number of elementary atoms in each compound atom. We have
already learned the rules framed by the founder of
the atomic theory for his guidance in attempting to
solve this problem. Berzelius thought those rules
insufficient and arbitrary ; he therefore laid down
two general rules, on the lines of which he prosecuted his researches into chemical synthesis.

"One atom of one element combines with one,
two, three, or more atoms of another element."
This is practically the same as Dalton's defini-
tions of binary, ternary, etc., compounds (p. 132).
"Two atoms of one element combine with three
and five atoms of another element." Berzelius here
recognizes the existence of compound atoms of a
more complex structure than any of those recognized by Dalton.

Berzelius further extended the conception of
atom by applying it to groups of elements formed,
according to him, by the combination of various
compound atoms. To his mind every compound

atom appeared as built up of two parts; each of these parts might be an elementary atom, or might be itself built up of several elementary atoms, yet in the Berzelian theory each acted as a definite whole. So far as the building up of the complex atom went, each of the two parts into which this atom could be divided acted as if it were a simple atom.

If we suppose a patch of two shades of red colour to be laid on a smooth surface, and alongside of this a patch of two shades of yellow colour, and if we suppose the whole mass of colour to be viewed from a distance such that one patch appears uniformly red and the other uniformly yellow, we shall have a rough illustration of the Berzelian compound atom. To the observer the whole mass of colour appears to consist of two distinct patches of contrasted colours; but let him approach nearer, and he perceives that what appeared to be a uniform surface of red or yellow really consists of two patches of unlike shades of red or of yellow. The whole mass of colour represents the compound atom; broadly it consists of two parts —the red colour represents one of the constituent atoms, the yellow colour represents the other constituent atom; but on closer examination the red atom, so to speak—and likewise the yellow atom— is found to consist of parts which are less unlike each other than the whole red atom is unlike the whole yellow atom.

We shall have to consider in more detail the reasoning whereby Berzelius arrived at this concep-

tion of every compound atom as a *dual* structure
(see pp. 209–212). At present I wish to notice this
conception as lying at the root of most of the work
which he did in extending and applying the Daltonian theory. I wish to insist on the fact that the
atomic theory could not advance without methods
being found for determining the number of elementary atoms in a compound atom, without clear
conceptions being gained of every compound atom
as a structure, and without at least attempts being
made to learn the laws in accordance with which
that structure was built. Before the atomic weight
of oxygen could be determined it was necessary
that the number of oxygen and of hydrogen atoms
in the atom of water should be known ; otherwise
all that could be stated was, the atomic weight of
oxygen is a simple multiple of 8. Berzelius did
much to advance chemical science by the introduction and application of a few simple rules whereby
he determined the number of elementary atoms in
various compound atoms. But as the science advanced, and as more facts came to be known, the
Berzelian rules were found to be too narrow and too
arbitrary ; chemists sought for some surer and more
generally applicable method than that which Berzelius had introduced, and the imperious demand for
this method at last forced them to recognize the
importance of the great generalization of the Italian
naturalist Avogadro, which they had possessed
since the year 1811, but the meaning of which they
had so long failed to understand.

Berzelius made one great step in the direction of recognizing Avogadro's distinction between atom and molecule when he accepted Gay-Lussac's generalization that "equal volumes of gases contain equal numbers of atoms:" but he refused to apply this to other than elementary gases. The weights of the volumes of elementary gases which combined were, for Berzelius, also the weights of the atoms of these elements. Thus, let the weight of one volume of hydrogen be called 1, then two volumes of hydrogen, weighing 2, combine with one volume of oxygen, weighing 16, to form two volumes of water vapour; therefore, said Berzelius, the atom of water consists of two atoms of hydrogen and one atom of oxygen, and the atom of the latter element is sixteen times heavier than the atom of the former. Three volumes of hydrogen, weighing 3, combine with one volume of nitrogen, weighing 14, to form two volumes of ammonia; therefore, said Berzelius, the atom of ammonia consists of three atoms of hydrogen combined with one atom of nitrogen, and the nitrogen atom is fourteen times heavier than the atom of hydrogen.

While Berzelius was applying these rules to the determination of the atomic weights of the elements, and was conducting the most important series of analyses known in the annals of the science, two great physico-chemical discoveries were announced.

In the year 1818 the "*law of isomorphism*"

was stated by Mitscherlich: "Compounds the atoms of which contain equal numbers of elementary atoms, similarly arranged, have the same crystalline form." As thus stated, the law of isomorphism affirms that if two compounds crystallize in the same form, the atoms of these compounds are built up of the same number of elementary atoms—however different may be the nature of the elements in the compounds—and that these elementary atoms are similarly arranged. This statement was soon found to be too absolute, and was accordingly modified; but to go into the history of the law of isomorphism would lead us too far from the great main path of chemical advance, the course of which we are seeking to trace.

Berzelius at once accepted Mitscherlich's law, as an aid in his researches on atomic weights. The help to be derived from this law may be illustrated thus: let us assume that two compounds have been obtained exhibiting identity of crystalline form; let it be further assumed that the number of elementary atoms in the atom of one of these compounds is known; it follows, by the law of isomorphism, that the number of elementary atoms in the atom of the other is known also. Let the two compounds be *sulphate of potash* and *chromate of potash;* let it be assumed that the atom of the first named is known to consist of two atoms of potassium, one atom of sulphur, and four atoms of oxygen; and that the

second substance is known to be a compound of the
elements potassium, chromium and oxygen; then
the atom of the second compound contains, by
Mitscherlich's law, two atoms of potassium, one
atom of chromium and four atoms of oxygen : hence
the relative weight of the atom of chromate of
potash can be determined, and hence the relative
weight of the atom of chromium can also be
determined.

A year after the announcement of Mitscherlich's
law, the following generalization was stated to hold
good, by two French naturalists, Dulong and Petit :
—" The atoms of all solid elements have the same
capacity for heat."

If the amount of heat required to raise the temperature of one grain of water through one degree
be called *one unit of heat*, then the capacity for
heat of any body other than water is the number
of units of heat required to raise the temperature
of one grain of that substance through one degree.
Each chemical substance, elementary and compound, has its own capacity for heat; but, instead
of comparing the capacities for heat of equal
weights, Dulong and Petit compared the capacities
for heat of weights representing the weights of the
atoms of various elements.   Thus, equal amounts of
heat are required to raise, through the same interval
of temperature, fifty-six grains of iron, one hundred
and eight grains of silver, and sixty-three and a half
grains of copper ; but the weights of the atoms of
these three elements are in the proportion of 56 :

108 : 63½. Dulong and Petit based their generalization on measurements of the capacities for heat of thirteen elements; further research has shown that their statement most probably holds good for all the solid elements. Here then was a most important instrument put into the hands of the chemist.

It is only necessary that the atomic weight of one solid element should be certainly known, and that the amount of heat required to raise through one degree the number of grains of that element expressed by its atomic weight should also be known ; then the number which expresses the weight, in grains, of any other solid element which is raised through one degree by the same amount of heat, likewise expresses the relative weight of the atom of that element. Thus, suppose that the atomic weight of silver is known to be 108, and suppose that six units of heat are required to raise the temperature of one hundred and eight grains of this metal through one degree ; then suppose it is found by experiment that six units of heat suffice to raise the temperature of two hundred and ten grains of bismuth through one degree, it follows—according to the law of Dulong and Petit—that 210 is the atomic weight of bismuth.

The modified generalization of Gay-Lussac— " Equal volumes of *elementary* gases contain equal numbers of atoms ; " the laws of " isomorphism " and of " atomic heat ; " and the two empirical rules stated on p. 163 ;—these were the guides used by

Berzelius in interpreting the analytical results which he and his pupils obtained in that memorable series of researches, whereby the conceptions of Dalton were shown to be applicable to a wide range of chemical phenomena.

The fixity of composition of chemical compounds has now been established; a definite meaning has been given to the term "element;" the conception of "atom" has been gained, but much remains to be done in the way of rendering this conception precise; and fairly good, but not altogether satisfactory methods have been introduced by which the relative weights of the atoms of elements and compounds may be determined. At this time chemists are busy preparing and describing new compounds, and many new elements are also being discovered; the need of classification begins to be felt more and more.

In the days of Berzelius and Davy strenuous efforts were made to obtain some generalizations by the application of which the many known elements and compounds might be divided into groups. It was felt that a classification might be founded on the composition of compounds, or perhaps on the properties of the same compounds. These two general principles served as guides in most of the researches then instituted; answers were sought to these two questions: Of what elements is this compound composed? and, What can this compound do; how does it react towards other bodies?

Lavoisier, as we know, regarded oxygen as the characteristic element of all *acids*. This term *acid* implies the possession, by all the substances denoted by it, of some common property; let us shortly trace the history of this word in chemistry.

Vinegar was known to the Greeks and Romans, and the names which they gave this substance tell us that sourness was to them its characteristic property. They knew that vinegar effervesced when brought into contact with chalky earths, and that it was able to dissolve many substances—witness the story of Cleopatra's draught of the pearl dissolved in vinegar. Other substances possessed of these properties—for instance oil of vitriol and spirits of salt—as they became known, were classed along with vinegar; but no attempts were made to clearly define the properties of these bodies till comparatively recent times.

The characteristics of an acid substance enumerated by Boyle are—solvent power, which is exerted unequally on different bodies; power of turning many vegetable blues to red, and of restoring many vegetable colours which had been destroyed by alkalis; power of precipitating solid sulphur from solutions of this substance in alkalis, and the power of acting on alkalis to produce substances without the properties of either acid or alkali.

But what, one may ask, is an alkali, of which mention is so often made by Boyle?

From very early times it had been noticed that the ashes which remained when certain plants were burned, and the liquid obtained by dissolving those ashes in water, had great cleansing powers; that they removed oily matter, fat and dirt from cloth and other fabrics. The fact that an aqueous solution of these ashes affects the coloured parts of many plants was also noticed in early times. As progress was made in chemical knowledge observers began to contrast the properties of this plant-ash with the properties of acids. The former had no marked taste, the latter were always very sour; the former turned some vegetable reds to blue, the latter turned the blues to red; a solution of plant-ash had no great solvent action on ordinary mineral matter, whereas this matter was generally dissolved by an acid. In the time of the alchemists, who were always seeking for the principles or essences of things, these properties of acids were attributed to *a principle of acidity*, while the properties of plant-ash and substances resembling plant-ash were attributed to a *principle of alkalinity* (from Arabic *alkali*, or *the ash*).

In the seventeenth century the distinction between acid and alkali was made the basis of a system of chemical medicine. The two principles of acidity and alkalinity were regarded as engaged in an active and never-ending warfare. Every disease was traced to an undue preponderance of one or other of these principles; to keep these unruly principles in quietness became the aim of the physician, and of course it was necessary that the physician should be a chemist, in order that he might know the nature and habits of the principles which gave him so much trouble.

Up to this time the term "alkali" had been applied to almost any substance having the properties which I have just enumerated ; but this group of substances was divided by Van Helmont and his successors into *fixed alkali* and *volatile alkali*, and fixed alkali was further subdivided into *mineral alkali* (what we now call soda) and *vegetable alkali* (potash). About the same time acids were likewise divided into three groups; *vegetable, animal,* and *mineral acids*. To the properties by which alkali was distinguished, viz. cleansing power and action on vegetable colouring matters, Stahl (the founder of the phlogistic theory) added that of combining with acids. When an acid (that is, a sour-tasting substance which dissolves most earthy matters and turns vegetable blues to red) is added to an alkali (that is, a substance which feels soap-like to the touch, which does not dissolve many earthy matters, and which turns many vegetable reds to blue) the properties of both acid and alkali disappear, and a new substance is produced which is not characterized by the properties of either constituent. The new substance, as a rule, is without action on earthy matters or on vegetable colours ; it is not sour, nor is it soapy to the touch like alkali ; it is *neutral*. It is *a salt*. But, although Stahl stated that an alkali is a substance which combines with an acid, it was not until a century later that these three—alkali, acid, salt—were clearly distinguished.

But the knowledge that a certain group of bodies are sour and dissolve minerals, etc., and that a certain other group of bodies are nearly tasteless and do not dissolve minerals, etc., was evidently a knowledge of only the outlying properties of the bodies ; it simply enabled a term to be applied to a group of bodies, which term had a definite connotation.

Why are acids acid, and *why* are alkalis alkaline?

Acids are acid, said Becher (latter part of seventeenth century), because they all contain the same principle, viz. the primordial acid. This primordial acid is more or less mixed with earthy matter in all actual acids ; it is very pure in spirits of salt.

Alkalis are alkaline, said Basil Valentine (beginning of the sixteenth century), because they contain a special kind of matter, " the matter of fire."

According to other chemists (*e.g.* J. F. Meyer, 1764), acids owe their acidity to the presence of a sharp or biting principle got from fire.

Acids, alkalis and salts *all* contain, according to Stahl (beginning of the eighteenth century), more or less *primordial acid*. The more of this a substance contains, the more acid it is ; the less of this it contains, the more alkaline it is.

All these attempted explanations recognize that similar properties are to be traced to similarity of composition ; but the assertion of the existence of a " primordial acid," or of " the matter of fire," although undoubtedly a step in advance, was not sufficiently definite (unless it was supplemented by a distinct account of the properties of these principles) to be accepted when chemical knowledge became accurate.

The same general consideration, founded on a large accumulation of facts, viz. that similarity of properties is due to similarity of composition, guided Lavoisier in his work on acids. He found the "primordial acid" of Stahl, and the "biting principle" of Meyer, in the element oxygen.

I have already (p. 91) shortly traced the reasoning whereby Lavoisier arrived at the conclusion that oxygen is *the acid-producer;* here I would insist on the difference between his method and that of Basil Valentine, Stahl and the older chemists. *They* carried into the domain of natural science conceptions obtained from, and essentially belonging to the domain of metaphysical or extra-physical speculation; *he* said that oxygen is the acidifier, because all the compounds of this element which he actually examined were pos_ sessed of the properties included under the name acid. We know that Lavoisier's conclusion was erroneous, that it was not founded on a sufficiently broad basis of facts. The conception of an acidifying principle, although that principle was identified with a known element, was still tainted with the vices of the alchemical school. We shall see immediately how much harm was done by the assertion of Lavoisier, "All acids contain oxygen."

In Chapter II. (pp. 32–37) we traced the progress of knowledge regarding alkalis from the time when the properties of these bodies were said to be due to the existence in them of " matter of fire," to the time when Black had clearly distinguished and defined caustic alkali and carbonated alkali.

The truly philosophical character, and at the same time the want of enthusiasm, of Black become apparent if we contrast his work on alkali with that of Lavoisier on acid. Black did not hamper the advance of chemistry by finding a " principle of alkalinity ;" but neither did he give a full explanation of the fact that certain bodies are alkaline while others are not. He set himself the problem of accurately determining the differences in composition between burnt (or caustic) and unburnt (or mild) alkali, and he solved the problem most successfully. He showed that the properties of mild alkalis differ from those of caustic alkalis, because the composition of the former differs from that of the latter ; and he showed exactly wherein this difference of composition consists, viz. in the possession or non-possession of fixed air.

Strange we may say that this discovery did not induce Black to prosecute the study of caustic alkalis : surely he would have anticipated Davy, and have been known as the discoverer of potassium and sodium.

In the time of Stahl the name "salt" was applied, as we have learned, to the substance produced by the union of an acid with an alkali ; but the same word was used by the alchemists with an altogether different signification. Originally applied to the solid matter obtained by boiling down sea-water, and then extended to include all substances which, like this solid matter, are very easily dissolved by water and can be recovered by boiling down this solution, "salt" was, in the sixteenth and seventeenth centuries, the name given to one of the hypothetical principles or elements. Many kinds of matter were known to be easily dissolved by water; the common possession of these properties was sought to be accounted for by saying that all these substances contained the same principle, namely, *the principle of salt.* I have already tried to indicate the reasoning whereby Boyle did so much to overthrow this conception of salt. He also extended our knowledge of special substances which are now classed as salts. The chemists who came after Boyle gradually reverted to the older meaning of the term "salt," adopting as the characteristics of all substances placed in this class, ready solubility in water, fusibility, or sometimes volatility, and the possession of a taste more or less like that of sea-salt.

Substances which resembled salts in general appearance, but were insoluble in water, and very fixed in the fire, were called "earths"; and, as was generally done in those days, the existence of a primordial earth was assumed, more or less of which was supposed to be present in actual earths. This recognition of the possibility of more or less of the primordial earth being present in actually occurring earths, of course necessitated the existence of

III. N

various kinds of earth. The earths were gradually distinguished from each other; lime was recognized as a substance distinct from baryta, baryta as distinct from alumina, etc.

Stahl taught that one essential property of an earth was fusibility by fire, with production of a substance more or less like glass. This property was possessed in a remarkable degree by quartz or silica. Hence silica was regarded as the typical earth, until Berzelius, in 1815, proved it to be an acid. But the earths resembled alkalis, inasmuch as they too combined with, and so neutralized, acids.

There is an alkali hidden in every earth, said some chemists.

An alkali is an earth refined by the presence of acid and combustible matter, said others.

Earths thus came to be included in the term "alkali," when that term was used in its widest acceptation. But a little later it was found that some of the earths were thrown down in the solid form from their solutions in acids by the addition of alkalis; this led to a threefold division, thus—

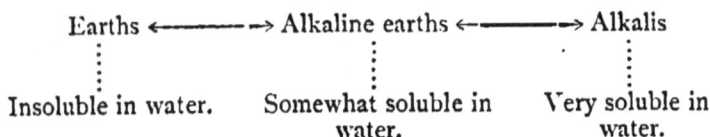

Earths ←————→ Alkaline earths ←————→ Alkalis

Insoluble in water.   Somewhat soluble in   Very soluble in
                              water.   water.

The distinction at first drawn between "earth" and "alkali" was too absolute; the intermediate group of "alkaline earths" served to bridge over the gap between the extreme groups.

"In Nature," says Wordsworth, "everything is

distinct, but nothing defined into absolute independent singleness."

At this stage of advance, then, an earth is regarded as differing from an alkali in being insoluble, or nearly insoluble in water; in not being
soapy to the touch, and not turning vegetable reds
to blue: but as resembling an alkali, in that it
combines with and neutralizes an acid; and the
product of this neutralization, whether accomplished
by an alkali or by an earth, is called a salt. To
the earth or alkali, as being the foundation on
which the salt is built, by the addition of acid,
the name of *base* was given by Rouelle in 1744.

But running through every conception which was
formed of these substances—acid, alkali, earth, salt
—we find a tendency, sometimes forcibly marked,
sometimes feebly indicated, but always present, to
consider salt as a term of much wider acceptation
than any of the others. An acid and an alkali, or
an acid and an earth, combine to form a salt; but
the salt could not have been thus produced unless
the acid, the alkali and the earth had contained
in themselves some properties which, when combined, form the properties of the salt.

The acid, the alkali, the earth, each is, in a
sense, a salt. The perfect salt is produced by the
coalescence of the saltness of the acid with the
saltness of the alkali. This conception finds full
utterance in the names, once in common use, of *sal
acidum* for acid, *sal alkali* for alkali, and *sal salsum*
or *sal neutrum* for salt. All are salts; at one extreme

comes that salt which is marked by properties called acid properties, at the other extreme comes the salt distinguished by alkaline properties, and between these, and formed by the union of these, comes the middle or neutral salt.

It is thus that the nomenclature of chemistry marks the advances made in the science. "What's in a name?" To the historical student of science, almost everything.

We shall find how different is the meaning attached in modern chemistry to these terms, *acid salt, alkaline salt, neutral salt*, from that which our predecessors gave to their *sal acidum, sal alkali*, and *sal neutrum*.

We must note the appearance of the term *vitriol*, applied to the solid salt-like bodies obtained from acids and characterized by a glassy lustre. By the middle of last century the vitriols were recognized as all derived from, or compounded of, sulphuric acid (oil of vitriol) and metals; this led to a sub-division of the large class of neutral salts into (1) metallic salts produced by the action of sulphuric acid on metals, and (2) neutral salts produced by the action of earths or alkalis on acids generally.

To Rouelle, a predecessor of Lavoisier, who died four years before the discovery of oxygen, we owe many accurate and suggestive remarks and experiments bearing on the term "salt." I have already mentioned that it was he who applied the word "base" to the alkali or earth, or it might be metal, from which, by the action of acid, a salt is built up. He also ceased to speak of an acid as *sal acidum*, or of an alkali as *sal alkali*, and applied the term " salt " exclusively to those substances which are produced by the action of acids on bases. When the product of such an action was neutral—that is, had no sour taste, no soapy feeling to the touch, no action on vegetable colours, and no action on acids or bases—he called that product *a neutral salt;* when the product still exhibited some of the properties of acid, *e.g.* sourness of taste, he called it *an acid salt;* and when the product continued to exhibit some of the properties of alkali, *e.g.* turned vegetable reds to blue, he called it *an alkaline salt.*

Rouelle also proved experimentally that an acid salt contains more acid—relatively to the same amount of base—than a neutral salt, and that an alkaline salt contains more base—relatively to the same amount of acid—than a neutral salt ; and he proved that this excess of acid, or of base, is chemically united to the rest of the salt—is, in other words, an essential part of the salt, from which it cannot be removed without changing the properties of the whole.

But we have not as yet got to know why certain qualities connoted by the term " acid " can be affirmed to belong to a group of bodies, why certain other, " alkaline," properties belong to another group, nor why a third group can be distinguished from both of these by the possession of properties which we sum up in the term " earthy." Surely

there must be some peculiarity in the composition
of these substances, common to all, by virtue of
which all are acid. The atom of an acid is
surely composed of certain elements which are
never found in the atom of an alkali or an earth ;
or perhaps the difference lies in the number, rather
than in the nature of the elements in the acid
atoms, or even in the arrangement of the elementary atoms in the compound atom of acid, of
alkali, and of earth.

I think that our knowledge of salt is now more
complete than our knowledge of either acid, alkali,
or earth. We know that a salt is formed by the
union of an acid and an alkali or earth ; if, then,
we get to know the composition of acids and bases
(*i.e.* alkalis and earths), we shall be well on the way
towards knowing the composition of salts.

And now we must resume our story where we
left it at p. 176. Lavoisier had recognized oxygen
as the acidifier ; Black had proved that a caustic
alkali does not contain carbonic acid.

Up to this time metallic calces, and for the most
part alkalis and earths also, had been regarded as
elementary substances. Lavoisier however proved
calces to be compounds of metals and oxygen ;
but as some of those calces had all the properties
which characterized earths, it seemed probable that
all earths are metallic oxides, and if all earths,
most likely all alkalis also. Many attempts were
made to decompose earths and alkalis, and to
obtain the metal, the oxide of which the earth or

the alkali was supposed to be. One chemist thought he had obtained a metal by heating the earth baryta with charcoal, but from the properties of his metal we know that he had not worked with a pure specimen of baryta, and that his supposed metallic base of baryta was simply a little iron or other metal, previously present in the baryta, or charcoal, or crucible which he employed.

But if Lavoisier's view were correct—if all bases contained oxygen—it followed that all salts are oxygen compounds. Acids all contain oxygen, said Lavoisier; this was soon regarded as one of the fundamental facts of chemistry. Earths and alkalis are probably oxides of metals; this before long became an article of faith with all orthodox chemists. Salts are produced by the union of acids and bases, therefore all salts contain oxygen: the conclusion was readily adopted by almost every one.

When the controversy between Lavoisier and the phlogistic chemists was at its height, the followers of Stahl had taunted Lavoisier with being unable to explain the production of hydrogen (or phlogiston as they thought) during the solution of metals in acids; but when Lavoisier learned the composition of water, he had an answer sufficient to quell these taunts. The metal, said Lavoisier, decomposes the water which is always present along with the acid, hydrogen is thus evolved, and the metallic calx or oxide so produced dissolves in the acid and forms a salt. If this explanation were correct—and there was an immense mass of evidence in its

favour and apparently none against it—then all
the salts produced by the action of acids on metals
necessarily contained oxygen.

The Lavoisierian view of a salt, as a compound
of a metallic oxide—or base—with a non-metallic
oxide—or acid—seemed the only explanation which
could be accepted by any reasonable chemist: in
the early years of this century it reigned supreme.

But even during the lifetime of its founder this
theory was opposed and opposed by the logic of
facts.   In 1787 Berthollet published an account of
experiments on prussic acid,—the existence and
preparation (from Prussian blue) of which acid had
been demonstrated three or four years before by the
Swedish chemist Scheele—which led him to conclude this compound to be a true acid, but free from
oxygen.   In 1796 the same chemist studied the
composition and properties of sulphuretted hydrogen, and pronounced this body to be an acid
containing no oxygen.

But the experiments and reasoning of Berthollet
were hidden by the masses of facts and the cogency
of argument of the Lavoisierian chemists.

The prevalent views regarding acids and bases
were greatly strengthened by the earlier researches
of Sir Humphry Davy, in which he employed the
voltaic battery as an instrument in chemical investigation.   Let us now consider some of the electro-
chemical work of this brilliant chemist.

In the spring of the year 1800 the electrical
battery, which had recently been discovered by

Volta, was applied by Nicholson and Carlisle to effect the decomposition of water. The experiments of these naturalists were repeated and confirmed by Davy, then resident at Bristol, who followed up this application of electricity to effect chemical changes by a series of experiments extending from 1800 to 1806, and culminating in the Bakerian Lecture delivered before the Royal Society in the latter year.

The history of Davy's life during these years, years rich in results of the utmost importance to. chemical science, will be traced in the sequel; meanwhile we are concerned only with the results of his chemical work.

The first Bakerian Lecture of Humphry Davy, "On some Chemical Agencies of Electricity," deserves the careful study of all who are interested in the methods of natural science ; it is a brilliant example of the disentanglement of a complex natural problem.

Volta and others had subjected water to the action of a current of electricity, and had noticed the appearance of acid and alkali at the oppositely electrified metallic surfaces. According to some experimenters, the acid was nitrous, according to others, muriatic acid. One chemist asserted the production of a new and peculiar body which he called *the electric acid.* The alkali was generally said to be ammonia.

When Davy passed an electric current through distilled water contained in glass vessels, connected

by pieces of moist bladder, cotton fibre, or other
vegetable matters, he found that nitric and hydrochloric acids were formed in the water surrounding
the positively electrified plate or pole, and soda
around the negatively electrified pole, of the
battery.

When the same piece of cotton fibre was re-
peatedly used for making connection between the
glass vessels, and was washed each time in dilute
nitric acid, Davy found that the production of
muriatic acid gradually ceased ; hence he traced the
formation of this acid to the presence of the animal
or vegetable substance used in the experiments.

Finding that the glass vessels were somewhat
corroded, and that the greater the amount of cor-
rosion the greater was the amount of soda making
its appearance around the negative pole, he concluded that the soda was probably a product of the
decomposition of the glass by the electric current ;
he therefore modified the experiment. He passed
an electric current through distilled water contained
in small cups of agate, previously cleaned by boiling in distilled water for several hours, and connected by threads of the mineral asbestos, chosen
as being quite free from vegetable matter; alkali
and acid were still produced. The experiment was
repeated several times with the same apparatus ;
acid and alkali were still produced, but the alkali
decreased each time. The only conclusion to be
drawn was that the alkali came from the water
employed. Two small cups of gold were now used

to contain the water; a very small amount of alkali appeared at the negative pole, and a little nitric acid at the positive pole. The quantity of acid slowly increased as the experiment continued, whereas the quantity of alkali remained the same as after a few minutes' action of the electric current. The production of alkali is probably due, said Davy, to the presence in the water of some substance which is not removed by distillation in a glass retort. By boiling down in a silver dish a quantity of the water he had used, a very small amount of solid matter was obtained, which after being heated was distinctly alkaline. Moreover when a little of this solid matter was added to the water contained in the two golden cups, there was a sudden and marked increase in the amount of alkali formed around the negative pole. Another quantity of the water which he had used was again distilled in a silver retort, and a little of the distillate was subjected to electrolysis as before. No alkali appeared. A little piece of glass was placed in the water; alkali quickly began to form. Davy thus conclusively proved that the alkali produced during the electrolysis (i.e. decomposition by the electric current) of water is not derived from the water itself, but from mineral impurities contained in the water, or in the vessel in which the water is placed during the experiment. But the production of nitric acid around the positive pole was yet to be accounted for.

Before further experiments could be made it was

necessary that Davy should form an hypothesis—
that he should mentally connect the appearance of
the nitric acid with some other phenomenon sufficient to produce this appearance; he could then
devise experiments which would determine whether
the connection supposed to exist between the two
phenomena really did exist or not.

Now, of the constituents of nitric acid—nitrogen,
hydrogen and oxygen—all except the first named
are present in pure water ; nitrogen is present in
large quantity in the ordinary atmosphere. It was
only necessary to assume that some of the hydrogen
and oxygen produced during the electrolysis of
water seized on and combined with some of the
nitrogen in the air which surrounded that water,
and the continual production of nitric acid during
the whole process of electrolysis was explained.

But how was this assumption to be proved or dis-
proved ? Davy adopted a method frequently made
use of in scientific investigations :—remove the
assumed cause of a phenomenon ; if the phenomenon
ceases to be produced, the assumed cause is probably the real cause. Davy surrounded the little
gold cups containing the water to be electrolysed with a glass jar which he connected with an
air-pump ; he exhausted most of the air from the
jar and then passed the electric current through
the water. Very little nitric acid appeared. He
now again took out most of the air from the glass
jar, admitted some hydrogen to supply its place,
and again pumped this out. This process he re-

peated two or three times and then passed the electric current. *No* acid appeared in the water. He admitted air into the glass vessel ; nitric acid began to be produced. Thus he proved that whenever air was present in contact with the water being electrolysed, nitric acid made its appearance, and when the air was wholly removed the acid ceased to be produced. As he had previously shown that the production of this acid was not to be traced to impurities in the water, to the nature of the vessel used to contain the water, or to the nature of the material of which the poles of the battery were composed, the conclusion was forced upon him that the production of nitric acid in the water, and the presence of ordinary air around the water invariably existed together ; that if one of these conditions was present, the other was also present—in other words, that one was the cause of the other.

The result of this exhaustive and brilliant piece of work is summed up by Davy in these words : " It seems evident then that water, chemically pure, is decomposed by electricity into gaseous matter alone, into oxygen and hydrogen."

From the effects of the electric current on glass, Davy argued that other earthy compounds would probably undergo change under similar conditions. He therefore had little cups of gypsum made, in which he placed pure water, and passed an electric current through the liquid. Lime was formed around the negative, and sulphuric acid around the positive pole. Using similar apparatus, he proved that the electric current decomposes very many minerals into an earthy or alkaline base and an acid.

Picturing to himself the little particles of a salt as being split by the electric current each into two smaller particles, one possessed of acid and the other of alkaline properties, Davy thought it might be possible to intercept the progress of these smaller particles, which he saw ever travelling towards the positive and negative poles of the battery. He accordingly connected these small glass vessels by threads of washed asbestos ; in one of the outer vessels he placed pure water, in the other an aqueous solution of sulphate of potash, and in the central vessel he placed ammonia. The negative pole of the battery being immersed in the sulphate of potash, and the positive pole in the water, it was necessary for the particles of sulphuric acid—produced by the decomposition of the sulphate of potash—to travel through the ammonia in the central vessel before they could find their way to the positive pole. Now, ammonia and sulphuric acid cannot exist in contact—they instantly combine to form sulphate of ammonia ; the sulphuric acid particles ought therefore to be arrested by the ammonia. But the sulphuric acid made its appearance at the positive pole just as if the central vessel had contained water. It seemed that the mutual attraction ordinarily exerted between sulphuric acid and ammonia was overcome by the action of the electric current. Ammonia would generally present an insuperable barrier to the progress of sulphuric acid, but the electrical energy appeared to force the acid particles over this barrier; they passed towards their goal as if nothing stood in their way.

Experiments are now multiplied by Davy, and the general conclusion drawn is that "Hydrogen, the alkaline substances, the metals and certain metallic oxides are attracted by negatively electrified metallic surfaces, and repelled by positively electrified metallic surfaces; and contrariwise, that oxygen and acid substances are attracted by positively electrified metallic surfaces, and repelled by negatively electrified metallic surfaces; and these attractive and repulsive forces are sufficiently energetic to destroy or suspend the usual operation of chemical affinity."*

To account for this apparent suspension of the ordinary chemical laws, Davy supposes that chemical compounds are continually decomposed and re-formed throughout the liquid which is subjected to the electrical action. Thus, in the experiment with water, ammonia and sulphate of potash, he supposes that the sulphuric acid and ammonia do combine in the central vessel to form sulphate of ammonia, but that this compound is again decomposed, by the electrical energy, into sulphuric acid —which passes on towards the positive pole—and ammonia—which remains in the central vessel—

---

* For an explanation of this expression, "chemical affinity," see p. 206, et seq.

ready to combine with more sulphuric acid as that comes travelling onwards from its source in the vessel containing sulphate of potash to its goal in the vessel containing water.

The eye of the philosopher had pierced beneath the apparent stability of the chemical systems which he studied. To his vision there appeared in those few drops of water and ammonia and sulphate of potash a never-ceasing conflict of contending forces ; there appeared a continual shattering and rebuilding of the particles of which the masses were composed. The whole was at rest, the parts were in motion ; the whole was constant in chemical composition, the composition of each particle was changed a thousand times in the minutest portion of every second. To the mind of Davy, the electrolysis of every chemical compound was a new application of the great law established by Newton —" To every action there is an equal and opposite reaction."

Each step made in chemical science since Davy's time has but served to emphasize the universality of this principle of action and reaction, a principle which has been too much overlooked in the chemical text-books, but the importance of which recent researches are beginning to impress on the minds of chemists.

It is the privilege of the philosophic student of Nature to penetrate the veil with which she conceals her secrets from the vulgar gaze. To him are shown sights which " eye hath not seen," and by him are perceived sounds which "ear hath not heard." Each drop of water is seen by him not only to be built up of myriads of small parts, but each particle is seen to be in motion ; many particles are being decomposed into still smaller particles of matter, different in properties from the original particles, but as the original particles are at the same time being reproduced, the continued existence of the drop of water with the properties of water is to him the result of the mutual action and reaction of contending forces. He knows that rest and permanence are gained, not by the cessation of action, but by the continuance of conflict ; he knows that in the realm of natural phenomena, stable equilibrium is the resultant of the action of opposite forces, and that complete decomposition occurs only when one force becomes too powerful or another becomes too weak.

Pursuing the train of thought initiated by the experiments which I have described, Davy entered upon a series of researches which led him to consider every chemical substance as possessing definite electrical relations towards every other substance. "As chemical attraction between two bodies seems to be destroyed by giving one of them an electrical state different from that which it naturally possessed—that is, by bringing it into a state similar to the other—so it may be increased by exalting its natural energy." Thus zinc, a metal easily oxidized, does not combine with oxygen when negatively electrified, whereas silver, a metal

III.   O

oxidized with difficulty, readily combines with oxygen when positively electrified.

Substances in opposite electrical states appear to combine chemically, and the greater the electrical difference the greater the readiness with which chemical combination is effected. Electrical energy and chemical attraction or *affinity* are evidently closely connected ; perhaps, said Davy, they are both results of the same cause.

Thus Davy arrived at the conception of a system of bodies as maintained in equilibrium by the mutual actions and reactions of both chemical and electrical forces; by increasing either of these a change is necessarily produced in the other. Under certain electrical conditions the bodies will exert no chemical action on one another, but such action may be started by changing these electrical conditions, or, on the other hand, by changes in the chemical relations of the bodies a change in the electrical relations may be induced. Thus Davy found that if plates of copper and sulphur are heated, the copper exhibits a positive and the sulphur a negative electrical condition ; that these electrical states become more marked as temperature rises, until the melting point of sulphur is reached, when the copper and sulphur combine together chemically and produce sulphide of copper.

When water is electrolysed, Davy looked on the oppositely electrified metallic plates in the battery as striving to attain a state of equilibrium ; the negatively electrified zinc strives to gain positive electricity from the copper, which strives to gain negative electricity from the zinc. The water he regarded as the carrier of these electricities, the one in this direction, the other in that. In thus acting as a carrier, the water is itself chemically decomposed, with production of hydrogen and oxygen ; but this chemical rearrangement of some of the substances which composed the original system (of battery and water) involves a fresh disturbance of electrical energy, and so the process proceeds until the whole of the water is decomposed or the whole of the copper or zinc plate is dissolved in the battery. If the water were not chemically decomposed, Davy thought that the zinc and copper in the battery would quickly attain the state of electrical equilibrium towards which they continually strive, and that the current would therefore quickly cease.

Davy thought that " however strong the natural electrical energies of the elements of bodies may be, yet there is every probability of a limit to their strength ; whereas the powers of our artificial instruments seem capable of indefinite increase." By making use of a very powerful battery, he hoped to be able to decompose substances generally regarded as simple bodies.

Taking a wide survey of natural phenomena, he sees these two forces, which we call chemical and electrical, everywhere at work, and by their mutual actions upholding the material universe in equilibrium. In the outbreaks of volcanoes he sees the disturbance of this equilibrium by the undue preponderance of electrical force; and in the formation of complex minerals beneath the surface of the earth, he traces the action of those chemical attractions which are ever ready to bring about the combination of elements, if they are not held in check by the opposing influence of electrical energy.

We shall see how the great and philosophical conception of Davy was used by Berzelius, and how, while undoubtedly gaining in precision, it lost much in breadth in being made the basis of a rigid system of chemical classification.

Davy's hope that the new instrument of research placed in the hands of chemists by Volta would be used in the decomposition of supposed simple substances was soon to be realized. A year after the lecture " On some Chemical Agencies of Electricity," Davy was again the reader of the Bakerian Lecture; this year (1807) it was entitled, "On some New Phenomena of Chemical Change produced by Electricity, particularly the Decomposition of the Fixed Alkalis; and the Exhibition of the New Substances which constitute their Bases; and on the General Nature of Alkaline Bodies."

In his first experiments on the effect of the electrical current on potash and soda, Davy used strong aqueous solutions of these alkalis, with the result that hydrogen and oxygen only were evolved. He then passed the current through melted potash kept liquid during the operation by the use of a spirit-lamp, the flame of which was fed with oxygen. Much light was evolved, and a great flame appeared at the negative pole ; on changing the direction of the current, "aeriform globules, which inflamed in the air, rose through the potash."

On the 6th of October 1807, a piece of potash was placed on a disc of platinum, which was made the negative pole of a very powerful battery; a platinum wire brought into contact with the upper surface of the potash served as the positive pole. When the current was passed, the potash became hot and soon melted ; gas was evolved at the upper surface, and at the lower (negative) side "there was no liberation of elastic fluid, but small globules, having a high metallic lustre, and being precisely similar in visible characters to quicksilver appeared, some of which burst with explosion and bright flame as soon as they were formed, and others remained, and were merely tarnished, and finally covered by a white film which formed on their surfaces."

When Davy saw these metallic globules burst through the crust of fusing potash, we are told by one of his biographers, "he could not contain his joy, he actually bounded about the room in ecstatic delight ; and some little time was required for him to compose himself sufficiently to continue the experiment."

This was the culminating point of the researches in which he had been continuously engaged for about six years. His interest and excitement were intense; the Bakerian Lecture was written "on the spur of the occasion, before the excitement of the mind had subsided," yet, says his biographer—and we may well agree with him—"yet it bears proof only of the maturest judgment; the greater part of it is as remarkable for experimental accuracy as for logical precision." But "to every action there is an equal and opposite reaction:" immediately after the delivery of the lecture, Davy was prostrated by a severe attack of illness, which confined him to bed for nine weeks, and was very nearly proving fatal.

That the phenomenon just described was really the decomposition of potash, and the production of the metal of which this substance is an oxygenized compound, was proved by obtaining similar results whether plates of silver, copper, or gold, or vessels of plumbago, or even charcoal, were used to contain the potash, or whether the experiment was conducted in the air, or in a glass vessel from which air had been exhausted, or in glass tubes wherein the potash was confined by mercury. The decomposition of potash was followed within a few days by that of soda, from which substance metallic globules were obtained which took fire when exposed to the air.

But the analysis of potash and soda was not sufficient for Davy; he determined to accomplish the synthesis of these substances. For this purpose he collected small quantities of the newly discovered metals, by conducting the electrolysis of potash and soda under experimental conditions such that the metals, as soon as produced, were plunged under the surface of naphtha, a liquid which does not contain oxygen, and which protected them from the action of the surrounding air.

A weighed quantity of each metal was then heated in a stream of pure dry oxygen, the products were collected and weighed, and it was found that solutions of these products in water possessed all the properties of aqueous solutions of potash and soda.

The new metals were now obtained in larger quantity by Davy, and their properties carefully determined by him; they were named *potassium* and *sodium* respectively. They were shown to possess all those properties which were generally accepted as characteristic of metal, except that of being heavy. The new metals were extremely light, lighter than water. For some time it was difficult to convince all chemists that a metal could be a very light substance. We are assured that a friend of Davy, who was shown potassium for the first time, and was asked what kind of substance he supposed it to be, replied, " It is metallic, to be sure ; " "and then, balancing it on his finger, he added in a tone of confidence, ' Bless me, how heavy it is ! ' "

Davy argued that since the alkalis, potash and soda, were found to be oxygen compounds of metals, the earths would probably also be found

to be metallic oxides. In the year 1808 he succeeded in decomposing the three earths, lime,
baryta and strontia, and in obtaining the metals
calcium, barium and *strontium*, but not in a perfectly
pure condition, or in any quantity. He also got
evidence of the decomposition of the earths silica,
alumina, zirconia and beryllia, by the action of
powerful electric currents, but he did not succeed
in obtaining the supposed metallic bases of these
substances.

So far Davy's discoveries had all tended to
confirm the generally accepted view which regarded
alkalis and earths as metallic oxides. But we
found that the outcome of these views was to
regard all salts—and among these, of course, common salt—as oxygen compounds.* Acids were
oxygen compounds, bases were oxygen compounds,
and as salts were produced by the union of acids
with bases, they, too, must necessarily be oxygen
compounds.

Berthollet had thrown doubt on the universality
of Lavoisier's name "oxygen," *the* acidifier, but he
had not conclusively proved the existence of any
acid which did not contain oxygen.

The researches of Davy naturally led him to
consider the prevalent views regarding acids, bases
and salts.

Muriatic (or as we now call it hydrochloric) acid
had long been a stumbling-block to the thorough-
going Lavoisierian chemists. Oxygen could not

* These views have been already explained on pp. 182, 183.

be detected in it, yet it ought to contain oxygen, because oxygen is the acidifier.   Of course, if muriatic acid contains oxygen, the salts—muriates— produced by the action of this acid on alkalis and earths must also contain oxygen.   'Many years before this time the action of muriatic acid on manganese ore had been studied by the Swedish chemist Scheele, who had thus obtained a yellow-coloured gas with a very strong smell.   Berthollet had shown that when a solution of this gas in water is exposed to sunlight, oxygen is evolved and muriatic acid is produced.   The yellow gas was therefore supposed to be, and was called, " oxidized muriatic acid," and muriatic acid was itself regarded as composed of oxygen and an unknown substance or *radicle.*

In 1809 Gay-Lussac and Thenard found that one volume of hydrogen united with one volume of the so-called oxidized muriatic acid to form muriatic acid ; the presence of hydrogen in this acid was therefore proved.

When Davy began (1810–11) to turn his at-tention specially to the study of salts, he adopted the generally accepted view that muriatic acid is a compound of oxygen and an unknown radicle, and that by the addition of oxygen to this com-pound oxidized muriatic acid is produced.   But unless Davy could prove the presence of oxygen in muriatic acid he could not long hold the opinion that oxygen was really a constituent of this substance.   He tried to obtain direct evidence

of the presence of oxygen, but failed. He then set
about comparing the action of muriatic acid on
metals and metallic oxides with the action of the
so-called oxidized muriatic acid on the same
substances. He showed that salt-like compounds
were produced by the action of oxidized muriatic
acid either on metals or on the oxides of these
metals, oxygen being evolved in the latter cases;
and that the same compounds and water were produced by the action of muriatic acid on the same
metallic oxides.

These results were most easily and readily explained by assuming the so-called oxidized muriatic
acid to be an elementary substance, and muriatic
acid to be a compound of this element with hydrogen. To the new element thus discovered—for he
who establishes the elementary nature of a substance may almost be regarded as its discoverer—
Davy gave the name of *chlorine*, suggested by the
yellow colour of the gas (from Greek, = *yellow*).
He at once began to study the analogies of chlorine,
to find by experiment which elements it resembled,
and so to classify it. Many metals, he found, combined readily with chlorine, with evolution of heat
and light. It acted, like oxygen, as a supporter of
combustion; it was, like oxygen, attracted towards
the negative pole of the voltaic battery; its compound with hydrogen was an acid; hence said
Davy chlorine, like oxygen, is a supporter of combustion and also an acidifier.

But it was very hard to get chemists to adopt

these views. As Bacon says, "If false facts in Nature be once on foot, what through neglect of examination, the countenance of antiquity, and the use made of them in discourse, they are scarce ever retracted."

Chemists had long been accustomed to systems which pretended to explain all chemical facts. The phlogistic theory, which had tyrannized over chemistry, had been succeeded by the Lavoisierian chemistry, which recognized one acidifier, and this also the one supporter of combustion. To ascribe these properties to any element other than oxygen appeared almost profane.

But when Davy spoke of chlorine as an acidifier, he did not use this word in the same sense as that in which it was employed by the upholders of the oxygen theory of acids; he simply meant to express the fact that a compound containing chlorine as one of its constituents, but not containing oxygen, was a true acid. When Gay-Lussac attempted to prove that hydrogen is an *alkalizing principle*, Davy said, "This is an attempt to introduce into chemistry a doctrine of occult qualities, and to refer to some mysterious and inexplicable energy what must depend upon a peculiar corpuscular arrangement." And with regard to Gay-Lussac's strained use of analogies between hydrogen compounds and alkalis, he says, "The substitution of analogy for fact is the bane of chemical philosophy; the legitimate use of analogy is to connect facts together, and to guide to new experiments."

But Davy's facts were so well established, and his experiments so convincing, that before two or three years had passed, most chemists were persuaded that chlorine was an element—*i.e.* a substance which had never been decomposed—and that muriatic acid was a compound of this element with hydrogen.

Berzelius was among the last to adopt the new view. Wöhler tells us that in the winter of 1823, when he was working in the laboratory of Berzelius, Anna, while washing some basins, remarked that they smelt strongly of oxidized muriatic acid: "Now," said Berzelius, "listen to me, Anna. Thou must no longer say 'oxidized muriatic acid,' but 'chlorine;' that is better."

This work on chlorine was followed up, in 1813, by the proof that the class of acidifiers and supporters of combustion contains a third elementary substance, viz. iodine. As Davy's views regarding acids and salts became developed, he seems to have more and more opposed the assumption that any one element is especially to be regarded as the acidifying element; but at the same time he seems to admit that most, if not all, acids contain hydrogen. Such oxides as sulphur trioxide, nitrogen pentoxide, etc., do not possess acid properties except in combination with water. But he of course did not say that all hydrogen compounds are acids; he rather regarded the possession by a substance of acid properties as dependent, to a great extent, on the nature of the elements other than hydrogen which it contained, or perhaps on the arrangement of all the elements in the particles of the acid. He regarded the hydrogen in an acid as capable of replacement by a metal, and to the metallic derivative—as it might be called—of the acid, thus produced, he gave the name of "salt." An acid might therefore be a compound of hydrogen with one other element—such were hydrochloric, hydriodic, hydrofluoric acids—or it might be a compound of hydrogen with two or more elements, of which one might or might not be oxygen—such were hydrocyanic acid and chloric or nitric acid. If the hydrogen in any of these acids were replaced by a metal a salt would be produced. A salt might therefore contain no oxygen, *e.g.* chloride or iodide of potassium ; but in most cases salts did contain oxygen, *e.g.* chlorate or nitrate of potassium.

Acids were thus divided into oxyacids (or acids which contain oxygen) and acids containing no oxygen ; the former class including most of the known acids. The old view of salts as being compounds of acids (*i.e.* oxides of the non-metallic elements) and bases (*i.e.* oxides of metals) was overthrown, and salts came to be regarded as metallic derivatives of acids.

From this time, these terms—acids, salts, bases —become of less importance than they formerly were in the history of chemical advance.

In trying to explain Davy's electro-chemical theory I have applied the word *affinity* to the mutual action and reaction between two substances which combine together to form a chemical compound. It is now necessary that we should look a little more closely into the history of this word *affinity*.

Oil and water do not mix together, but oil and potash solution do; the former may be said not to have, and the latter to have, an affinity one for the other. When sulphur is heated, the yellow odourless solid, seizing upon oxygen in the air, combines with it to produce a colourless strongly smelling gas. Sulphur and oxygen are said to have strong affinity for each other.

If equal weights of lime and magnesia be thrown into diluted nitric acid, after a time it is found that some of the lime, but very little of the magnesia, is dissolved. If an aqueous solution of lime be added to a solution of magnesia in nitric acid, the magnesia is precipitated in the form of an insoluble powder, while the lime remains dissolved in the acid. It is said that lime has a stronger affinity for nitric acid than magnesia has. Such reactions as these used to be cited as examples of *single elective affinity*—single, because one substance combined with one other, and elective, because a substance seemed to choose between two others presented to it, and to combine with one to the exclusion of the other.

But if a neutral solution of magnesia in sulphuric acid is added to a neutral solution of lime in nitric acid, sulphate of lime and nitrate of magnesia are produced. The lime, it was said, leaves the nitric and goes to the sulphuric acid, which, having been deserted by the magnesia, is ready to receive it; at the same time the nitric acid from which the lime has departed combines with the magnesia formerly held by the sulphuric acid. Such a reaction was said to be an instance of *double affinities*. The chemical changes were caused, it was said, by the simultaneous affinity of lime for sulphuric acid, which was greater than its affinity for nitric acid, and the affinity of magnesia for nitric acid, which was greater than its affinity for sulphuric acid.

If a number of salts were mixed, each base—supposing the foregoing statements to be correct—would form a compound with that acid for which it had the greatest affinity. It should then be possible to draw up tables of affinity. Such tables were indeed prepared. Here is an example:—

Sulphuric Acid.

Baryta.	Lime.
Strontia.	Ammonia.
Potash.	Magnesia.
Soda.	

This table tells us that the affinity of baryta for sulphuric acid is greater than that of strontia for the same acid, that of strontia greater than that of potash, and so on. It also tells that potash will decompose a compound of sulphuric acid and soda, just as soda will decompose a compound of the same acid with lime, or strontia will decompose a compound with potash, etc.

But Berthollet showed in the early years of this

century that a large quantity of a body having a weak affinity for another will suffice to decompose a small quantity of a compound of this other with a third body for which it has a strong affinity. He showed, that is, that the formation or non-formation of a compound is dependent not only on the so-called affinities between the constituents, but also on the relative quantities of these constituents. Berthollet and other chemists also showed that affinity is much conditioned by temperature ; that is, that two substances which show no tendency towards chemical union at a low temperature may combine when the temperature is raised. He, and they, also proved that the formation or non-formation of a compound is much influenced by its physical properties. Thus, if two substances are mixed in solution, and if by their mutual action a substance can be produced which is insoluble in the liquids present, that substance is generally produced whether the affinity between the original pair of substances be strong or weak.

The outcome of Berthollet's work was that tables of affinity became almost valueless. To say that the affinity of this body for that was greater than its affinity for a third body was going beyond the facts, because the formation of this or that compound depended on many conditions much more complex than those connoted by the term "affinity." Yet the conception of affinity remained, although it could not be applied in so rigorous a way as had been done by the earlier chemists. If an element,

A, readily combines with another element, B, under certain physical conditions, but does not, under the same conditions, combine with a third element, C, it may still be said that A and B have, and A and C have not, an affinity for each other.

This general conception of affinity was applied by Berzelius to the atoms of elements. Affinity, said Berzelius, acts between unlike atoms, and causes them to unite to form a compound atom, unlike either of the original atoms; cohesion, on the other hand, acts between like atoms, causing them to hold together without producing any change in their properties. Affinity varies in different elements. Thus the affinity of gold for oxygen is very small; hence it is that gold is found in the earth in the metallic state, while iron, having a great affinity for oxygen, soon rusts when exposed to air, or when buried in the earth. Potassium and sodium have great affinities for oxygen, chlorine, etc.; yet the atoms of potassium and sodium do not themselves combine. The more any elements are alike chemically the smaller is their affinity for each other; the more any elements are chemically unlike the greater is their mutual affinity; but this affinity is modified by circumstances. Thus, said Berzelius, if equal numbers of atoms of A and B, having equal or nearly equal affinity for C, mutually react, compound atoms, AC and BC, will be produced, but atoms of A and B will remain. The amounts of AC and BC produced will be influenced by the greater or less affinity of A and B for C;

but if there be a greater number of A than of B atoms, a greater amount of AC than of BC will be produced. In these cases all the reacting substances and the products of the actions are supposed to be liquids; but BC, if a solid substance, will be produced even if the affinity of A for C is greater than that of B for C.

In some elements, Berzelius taught, affinity slumbers, and can be awakened only by raising the temperature. Thus carbon in the form of coal has no affinity for oxygen at ordinary temperatures; it has remained for ages in the earth without undergoing oxidation; but when coal is heated the affinities of carbon are awakened, combination with oxygen occurs, and heat is produced.

But why is it that certain elementary atoms exhibit affinity for certain others? It depends, said Berzelius, on the electrical states of these atoms. According to the Berzelian theory, every elementary atom has attached to it a certain quantity of electricity, part of which is positive and part negative. This electricity is accumulated at two points on each atom, called respectively the positive pole and the negative pole; but in each atom one of these electricities so much preponderates over the other as to give the whole atom the character of either a positively or a negatively electrified body. When two atoms combine chemically the positive electricity in one neutralizes the negative electricity in the other. As we know that similar electricities repel, and opposite electricities attract

each other, it follows that a markedly positive atom
will exhibit strong affinity for a markedly negative
atom, less strong affinity for a feebly negative, and
little or no affinity for a positively electrified atom;
but two similarly electrified atoms may exhibit
affinity,. because in every positive atom there is
some negative electricity, as in every negative
atom there is some positive electricity. Thus,
in the atoms of copper and zinc positive electricity predominates, said Berzelius, but the zinc
atoms are more positive than those of copper;
hence, when the metals are brought into contact
the negative electricity of the copper atoms is
attracted and neutralized by the positive electricity of the zinc atoms, combination takes place,
and the compound atom is still characterized by a
predominance of positive electricity.

Hence Berzelius identified " electrical polarity "
with chemical affinity. Every atom was regarded
by him as *both* positively *and* negatively electrified ;
but as one of these electricities was always much
stronger than the other, every atom regarded as
a whole appeared to be *either* positively *or* negatively electrified. Positive atoms showed affinity
for negative atoms, and *vice versâ*. As a positive
atom might become more positive by increasing
the temperature of the atom, so might the affinity
of this atom for that be more marked at high than
at low temperatures.

Now, if two elementary atoms unite, the compound atom must—according to the Berzelian

views—be characterized either by positive or negative electricity. This compound atom, if positive, will exhibit affinity for other compound atoms in which negative electricity predominates ; if negative, it will exhibit affinity for other positively electrified compound atoms. If two compound atoms unite chemically, the complex atom so produced will, again, be characterized by one or other of the two electricities, and as it is positive or negative, so will it exhibit affinity for positively or negatively electrified complex atoms. Thus Berzelius and his followers regarded every compound atom, however complex, as essentially built up of two parts, one of which was positively and the other negatively electrified, and which were held together chemically by virtue of the mutual attractions of these electricities ; they regarded every compound atom as a *dual* structure. The classification adopted by Berzelius was essentially a dualistic classification. His system has always been known in chemistry as *dualism*.

Berzelius divided compound atoms (we should now say molecules) into three groups or orders—

Compound atoms of the first order, formed by the immediate combination of atoms of two, or in organic compounds of three, elementary substances.

Compound atoms of the second order, formed by the combination of atoms of an element with atoms of the first order, or by the combination of two or more atoms of the first order.

Compound atoms of the third order, formed by

combination of two or more atoms of the second order.

When an atom of the third order was decomposed by an electric current, it split up, according to the Berzelian teaching, into atoms of the second order—some positively, others negatively electrified. When an atom of the second order was submitted to electrolysis, it decomposed into atoms of the first order—some positively, others negatively electrified.

Berzelius said that a base is an electro-positive oxide, and an acid is an electro-negative oxide. The more markedly positive an oxide is, the more basic it is ; the more negative it is, the more is it characterized by acid properties.

One outcome of this teaching regarding acids and bases was to overthrow the Lavoisierian conception of oxygen as the acidifying element. Some oxides are positive, others negative, said Berzelius ; but acids are characterized by negative electricity, therefore the presence of oxygen in a compound does not always confer on that compound acid properties.

We have already seen that silica was regarded by most chemists as a typical earth ; but Berzelius found that in the electrolysis of compounds of silica, this substance appeared at the positive pole of the battery—that is, the atom of silica belonged to the negatively electrified order of atoms. Silica was almost certainly an oxide ; but electro-negative oxides are, as a class, acids ; therefore silica was

probably an acid. The supposition of the acid
character of silica was amply confirmed by the
mineralogical analyses and experiments of Berzelius. He showed that most of the earthy minerals
are compounds of silica with electro-positive metallic oxides, and that silica plays the part of an acid
in these minerals ; and in 1823 he obtained the
element silicon, the oxide of which is silica. On
this basis Berzelius reared a system of classification in mineralogy which much aided the advance
of that branch of natural science.

By the work of Berzelius and Davy the Lavoisierian conception of acid has now been much
modified and extended ; it has been rendered less
rigid, and is therefore more likely than before to be
a guide to fresh discoveries.

The older view of acid and alkali was based, for
the most part, on a qualitative study of the reactions of chemical substances: bodies were placed
in the same class because they were all sour, or
all turned vegetable blues to red, etc. This was
followed by a closer study of the composition of
substances, and by attempts to connect the properties of these substances with their composition ;
but when this attempt resulted in the promulgation
of the dictum that "oxygen is the acidifying principle," it began to be perceived that a larger basis
of fact must be laid before just conclusions could
be drawn as to the connections between properties
and composition of substances. This larger basis
was laid by the two chemists whose work we have

now reviewed. Of the life of one of these men I have already given such a sketch as I can from the materials available to me ; of the life of the other we happily possess ample knowledge. Let us now consider the main features of this life.

HUMPHRY DAVY, the eldest son of Robert and Grace Davy, was born at Penzance, in Cornwall, on December 17, 1778, eight months that is before the birth of Berzelius. His parents resided on a small property which had belonged to their ancestors for several generations. Surrounded by many kind friends by whom he was much thought of, the boy appears to have passed a very happy childhood. Even at the age of five his quickness and penetration were marked by those around him, and at school these continued to be his predominant characteristics. Nurtured from his infancy in the midst of beautiful and romantic scenery, and endowed with great observing power and a lively imagination, young Davy seemed destined to be one of those from whose lips is " poured the deathless singing ; " all through life he was characterized by a strongly marked poetic temperament.

Humphry Davy was held in much esteem by his school friends as a composer of valentines and love letters, as a daring and entertaining teller of stories, and as a successful fireworks manufacturer. Such a combination of qualities would much endear him to his boy-companions. We are told that at the age of eight he used to mount on an empty cart, around

which a circle of boys would collect to be entertained by the wonderful tales of the youthful
narrator.

Finishing his school education at the age of
fifteen, he now began his own education of himself.
In 1795 he was apprenticed to a surgeon and
apothecary (afterwards a physician), in Penzance,
with whom he learned the elements of medical
science ; but his time during the years which he
spent under Mr. Borlase was much occupied in
shooting, fishing, searching for minerals and geological specimens, composing poetry, and pursuing
metaphysical speculations. He was now, as through
life, an enthusiastic lover of Nature ; his mind was
extremely active, ranging over the most diverse
subjects ; he was full of imagination, and seemed
certain to distinguish himself in any pursuit to
which he should turn his attention. During the
next three or four years Davy indulged freely in
speculations in all manner of subjects ; he started,
as people generally do when young, from general
principles and followed these out to many conclusions. Even in his study of physiology and other
branches of science, he appears at this time to
have adopted the speculative rather than the experimental method ; but unlike most youthful
metaphysicians he was ready to give up an opinion
whenever it appeared to him incorrect. By the
time he reached the age of twenty he had dis-
carded this method of seeking for truth, and was
ever afterwards distinguished by his careful work-

ing out of facts as the foundation for all his brilliant theories.

Davy appears to have begun the study of
chemistry about 1798 by reading Lavoisier's
" Elements of Chemistry," the teachings of which
he freely criticized. About this time Mr. Gre-
gory Watt came to live at Penzance as a lodger
with Davy's mother, and with him the young philosopher had much talk on chemical and other scientific subjects. He also became acquainted with Mr.
Davies Gilbert—who was destined to succeed Davy
as President of the Royal Society—and from him
he borrowed books and received assistance of
various kinds in his studies.

It was during these years that Davy made experiments on heat, which were published some
years later, and which are now regarded as laying
the foundations of the modern theory according to
which heat is due to the motions of the small parts
of bodies. He arranged two brass plates so that one
should carry a block of ice which might be caused
to revolve in contact with the other plate; the
plates were covered by a glass jar, from which he
exhausted the air by means of a simple syringe of
his own contrivance; the machine being placed on
blocks of ice the plates were caused to revolve. The
ice inside the jar soon melted; Davy concluded
that the heat required to melt this ice could only
be produced by the friction of the ice and brass,
and that therefore heat could not be any form
of ponderable matter.

In the year 1798 Davy was asked to go to Bristol as superintendent of the laboratory of a new Pneumatic Institution started by Dr. Beddoes for the application of gases to the treatment of diseases. Davy had corresponded with Beddoes before this time regarding his experiments on heat, and the latter seems to have been struck with his great abilities and to have been anxious to secure him as experimenter for his institution. Davy was released from his engagements with Mr. Borlase, and, now about twenty years of age, set out for his new home, having made as he says all the experiments he could at Penzance, and eagerly looking forward to the better appliances and incitements to research which he hoped to find at Bristol.

The Pneumatic Institution was supported by subscriptions, for the most part from scientific men. It was started on a scientific basis. Researches were to be made on gases of various kinds with the view of applying these as remedies in the alleviation of disease. An hospital for patients, a laboratory for experimental research, and a lecture theatre were provided.

At this time many men of literary and intellectual eminence resided in Bristol; among these were Coleridge and Southey. Most of these men were visitors at the house of Dr. Beddoes, and many distinguished men came from various parts of the county to visit the institution. Davy thus entered on a sphere of labour eminently suited for the development of his genius. With ample mechanical appliances for research, with plenty of time at his disposal, surrounded by an atmosphere of inquiry and by men who would welcome any additions he could make to the knowledge of Nature, and being at the same time not without poetic and imaginative surroundings, by which he was ever spurred onwards in the pursuit of truth—placed in these circumstances, such an enthusiastic and diligent student of science as Davy could not but obtain results of value to his fellows.   The state of chemical science at this time was evidently such as to incite the youthful worker.   The chains with which Stahl and his successors had so long bound the limbs of the young science had been broken by Lavoisier ; and although the French school of chemistry was at this time dominant, and not disinclined to treat as ignorant any persons who might differ from its teaching, yet there was plenty of life in the cultivators of chemistry.   The controversy between Berthollet and Proust was about to begin ; the Lavoisierian views regarding acids and salts were not altogether accepted by Gay-Lussac, Thenard and others ; and from the laboratory of Berzelius there was soon to issue the first of those numerous researches which drew the attention of every chemist to the capital of Sweden.   The voltaic battery had been discovered, and had opened up a region of possibilities in chemistry.

Davy began his researches at the institution by experiments with nitrous oxide, a gas supposed by some people at that time to be capable of produc-

ing most harmful effects on the animal system. He
had to make many experiments before he found
a method for preparing the pure gas, and in the
course of these experiments he added much to the
stock of chemical knowledge regarding the compounds of nitrogen and oxygen.   Having obtained
fairly pure nitrous oxide, he breathed it from a silk
bag; he experienced a "sensation analogous to
gentle pressure on all the muscles; . . . the objects
around me became dazzling and my hearing more
acute; . . . at last an irresistible propensity to
action was indulged in. . . . I recollect but indis-
tinctly what followed; I know that my motions
were various and violent."   Southey and Coleridge
breathed the gas; the poets only laughed a little.
Encouraged by the results of these experiments,
Davy proceeded to prepare and breathe nitric
oxide—whereby he was rendered very ill—and
then carburetted hydrogen—which nearly killed
him.

In his chemical note-book about this time, Davy
says, "The perfection of chemical philosophy, or
the laws of corpuscular motion, must depend on
the knowledge of all the simple substances, their
mutual attractions, and the ratio in which the
attractions increase or diminish with increase or
diminution of temperature. . . . The first step towards these laws will be the decomposition of those
bodies which are at present undecompounded."
And in the same note-book he suggests methods
which he thinks might effect the decomposition of

muriatic and boric acids, the alkalis and earths. Here are the germs of his future work.

After about eight months' work at Bristol he published a volume of "Researches," which contained a great many new facts, and was characterized by vigour and novelty of conception. These researches had been carried out with intense application; each was struck off at a red heat. His mind during this time was filled with vast scientific conceptions, and he began also to think of fame. "An active mind, a deep ideal feeling of good, and a look towards future greatness," he tells us, sustained him.

Count Rumford, the founder of the Royal Institution in London, was anxious to obtain a lecturer on chemistry for the Institution. Davy was strongly recommended, and after a little arrangement—concerning which Davy says in a letter, " I will accept of no appointment except on the sacred terms of independence "—he was appointed Assistant Lecturer on Chemistry and Director of the Laboratory. About a year later his official designation was changed to Professor of Chemistry. This appointment opened up a great sphere of research; "the sole and uncontrolled use of the apparatus of the institution for private experiments " was to be granted him, and he was promised "any apparatus he might need for new experiments."

He had now the command of a good laboratory; he had not to undergo the drudgery of systematic teaching, but was only required to give lectures to a general audience. Before leaving Bristol he had commenced experiments on the chemical applications of the voltaic battery; these he at once followed up with the better apparatus now at his command. The results of this research, and his subsequent work on the alkalis and on muriatic acid and chlorine, have been already described. The circumstances of Davy's life had hitherto been most favourable; how nobly he had availed himself of these circumstances was testified by the work done by him.

His first lecture was delivered in the spring of 1801, and at once he became famous. A friend of Davy says, "The sensation created by his first course of lectures at the Institution, and the enthusiastic admiration which they obtained, is scarcely to be imagined. Men of the first rank and talent, the literary and the scientific, the practical and the theoretical, blue-stockings and women of fashion, the old and the young—all crowded, eagerly crowded the lecture-room. His youth, his simplicity, his natural eloquence, his chemical knowledge, his happy illustrations and well-conducted experiments, excited universal attention and unbounded applause. Compliments, invitations and presents were showered upon him in abundance from all quarters; his society was courted by all, and all appeared proud of his acquaintance." One of his biographers says of these lectures, "He was always in earnest, and when he amused most, amusement appeared most foreign to his object. His great and first object was to instruct, and in conjunction with this, maintain the importance and dignity of science; indeed, the latter, and the kindling a taste for scientific pursuits, might rather be considered his main object, and the conveying instruction a secondary one."

The greatest pains were taken by Davy in the composition and rehearsal of his lectures, and in the arrangement of experiments, that everything should tend towards the enlightenment of his audience. Surrounded by a brilliant society, invited to every fashionable entertainment, flattered by admirers, tempted by hopes of making money, Davy remained a faithful and enthusiastic student of Nature. "I am a lover of Nature," he writes at this time to a friend, "with an ungratified imagination. I shall continue to search for untasted charms, for hidden beauties. My *real*, my *waking* existence, is amongst the objects of scientific research. Common amusements and enjoyments are necessary to me only as dreams to interrupt the flow of thoughts too nearly analogous to enlighten and vivify."

During these years (*i.e.* from 1802 to 1812) he worked for the greater part of each day in the laboratory. Every week, almost every day, saw some fresh discovery of importance. He advanced from discovery to discovery. His work was characterized by that vast industry and extreme rapidity which belong only to the efforts of genius. Never, before or since, has chemical science made such strides in this country.

In 1803 Davy was elected a Fellow, and in 1807 one of the secretaries of the Royal Society. In 1812 he retired from the professorship of chemistry at the Royal Institution ; in the same year he was made a knight.

The next two or three years were mostly spent in travelling abroad with his wife—he had married a widow lady, Mrs. Apreece, in 1812. During his visit to Paris he made several experiments on the then recently discovered iodine, and proved this substance to be an element.

The work which Davy had accomplished in the seventeen years that had now elapsed since he began the study of chemistry, whether we consider it simply as a contribution to chemical science, or in the light of the influence it exerted on the researches of others, was of first-rate importance ; but a fresh field now began to open before him, from which he was destined to reap the richest fruits. In the autumn of 1815 his attention was drawn to the subject of fire-damp in coal-mines. As he passed through Newcastle, on his return from a holiday spent in the Scottish Highlands, he examined various coal-mines and collected samples of fire-damp ; in December of the same year his *safety-lamp* was perfected, and soon after this it was in the hands of the miner.

The steps in the discovery of this valuable instrument were briefly these. Davy established the fact that fire-damp is a compound of carbon and hydrogen ; he found that this gas must be mixed

with a large quantity of ordinary air before the
mixture becomes explosive, that the temperature
at which this explosion occurs is a high one,
and that but little heat is produced during the
explosion; he found that the explosive mixture
could not be fired in narrow metallic tubes, and
also that it was rendered non-explosive by addition of carbonic acid or nitrogen.   He reasoned
on these facts thus: "It occurred to me, as a
considerable heat was required for the inflammation of the fire-damp, and as it produced in
burning a comparatively *small degree* of heat,
that the effect of carbonic acid and azote, and
of the surfaces of small tubes, in preventing its
explosion, depended on their cooling powers—
upon their lowering the temperature of the exploding mixture so much that it was no longer
sufficient for its continuous inflammation."   He at
once set about constructing a lamp in which it
should be impossible for the temperature of ignition
of a mixture of fire-damp and air to be attained,
and which therefore, while burning, might be filled
with this mixture without any danger of an explosion.   He surrounded the flame of an oil-lamp
with a cylinder of fine wire-gauze ; this lamp when
brought into an atmosphere containing fire-damp
and air could not cause an explosion, because
although small explosions might occur in the
interior of the wire cylinder, so much heat was
conducted away by the large metallic surface that
the temperature of the explosive atmosphere out-

III.   Q

side the lamp could not attain that point at which explosion would occur.

In 1818 Sir Humphry Davy was made a baronet, in recognition of his great services as the inventor of the safety-lamp; and in 1820 he was elected to the most honourable position which can be held by a man of science in this country, he became the President of the Royal Society.

For seven years he was annually re-elected president, and during that time he was the central figure in the scientific society of England. During these years he continued his investigations chiefly on electro-chemical subjects and on various branches of applied science. In 1826 his health began to fail. An attack of paralysis in that year obliged him to relinquish most of his work. He went abroad and travelled in Italy and the Tyrol, sometimes strong enough to shoot or fish a little, or even to carry on electrical experiments; sometimes confined to his room, or to gentle exercise only. He resigned the presidentship of the Royal Society in 1827. In 1828 he visited Rome, where he was again attacked by paralysis, and thought himself dying, but he recovered sufficiently to attempt the journey homeward. At Geneva he became very ill, and expired in that city on the 29th of May 1829.

During these later years of illness and suffering, his intense love of and delight in Nature were very apparent; he returned again to the simple tastes and pleasures of his early days. His intimate knowledge of natural appearances and of the sights and sounds of country life is conspicuous in the "Salmonia, or Days of Fly-fishing," written during his later years.

Sir Humphry Davy was emphatically a genius. He was full of eager desire to know the secrets of the world in which he lived; he looked around him with wonder and delight, ever conscious of the vastness of the appearances which met his gaze; an exuberance of life and energy marked his actions; difficulties were encountered by him only to be overcome; he was depressed by no misfortunes, deterred by no obstacles, led aside from his object by no temptations, and held in bondage by no false analogies.

His work must ever remain as a model to the student of science. A thorough and careful foundation of fact is laid; on this, hypotheses are raised, to be tested first by reasoning and argument, then by the tests of the laboratory, which alone are final. Analogies are seized; hints are eagerly taken up, examined, and acted on or dismissed. As he works in the laboratory, we see his mind ranging over the whole field of chemical knowledge, finding a solution of a difficulty here, or guessing at a solution there; combining apparently most diverse facts; examining phenomena which appear to have no connection; never dwelling too long on an hypothesis which cannot yield some clue to the object of research, but quickly discovering the road which will lead to the wished-for solution.

Like so many great experimenters Davy accomplished wonders with little apparatus. When he went abroad for the first time he took with him two small boxes, one twenty, and the other twelve inches long, by about seven inches wide and four deep. With the apparatus contained in these boxes he established the elementary nature of iodine, and made a rough estimation of its atomic weight; he determined many of its analogies with chlorine, proving that, like chlorine, it is markedly electro-negative, and that its compounds are decomposed by chlorine; he accomplished the synthesis of hydriodic acid, and approximately determined the composition of iodide of nitrogen. But when it was necessary to employ delicate or powerful apparatus, he was able by the use of that also to obtain results of primary importance. The decomposition of potash, soda, baryta, lime and strontia could not have been effected had he not had at his command the resources of a well-furnished laboratory.

Davy has had no successor in England. Much useful and some brilliant work has been done by English chemists since his day, but we still look back to the first quarter of the century as the golden age of chemistry in this country. On the roll wherein are written the names of England's greatest sons, there is inscribed but a single chemist—Humphry Davy.

I carried on the account of the work of Davy's great contemporary, Berzelius, to the time when he had fairly established dualistic views of the structure of chemical compounds, and when, by the application of a few simple rules regarding the combinations of elementary atoms, he had largely extended the bounds of the atomic theory of Dalton.

Berzelius also did important work in the domain of organic chemistry. By numerous analyses of compounds of animal and vegetable origin, he clearly established the fact that the same laws of combination, the same fixity of composition, and the same general features of atomic structure prevail among the so-called organic as among the inorganic compounds. In doing this he broke down the artificial barrier which had been raised between the two branches of the science, and so prepared the way for modern chemistry, which has won its chief triumphs in the examination of organic compounds.

By the many and great improvements which he introduced into analytical chemistry, and by the publication of his " Textbook of Chemistry," which went through several editions in French and German, and also of his yearly report on the advance of chemistry, Berzelius exerted a great influence on the progress of his favourite science. Wöhler tells us that when the spring of the year came, at which time his annual report had to be prepared, Berzelius shut himself up in his study, surrounded himself with books, and did not stir from the writing-table until the work was done.

In his later days Berzelius was much engaged in controversy with the leaders of the new school, the rise and progress of which will be traced in the next chapter, but throughout this controversy he found time to add many fresh facts to those already known. He continued his researches until his death in 1848.

The work of the great Swedish chemist is characterized by thoroughness in all its parts: to him every fact appeared to be of importance; although now perhaps only an isolated fact, he saw that some day it would find a place in a general scheme of classification. He worked in great measure on the lines laid down by Dalton and Davy; the enormous number and accuracy of his analyses established the law of multiple proportions on a sure basis, and his attempts to determine the constitution of compound atoms, while advancing the atomic theory of Dalton, drew attention to the all-important distinction between atom and molecule, and so prepared chemists for the acceptance of the generalization of Avogadro. The electro-chemical conceptions of Davy were modified by Berzelius; they were shorn of something of their elasticity, but were rendered more suited to be the basis of a rigid theory.

At the close of this transition period from the Lavoisierian to the modern chemistry, we find analytical chemistry established as an art; we find the atomic theory generally accepted, but we notice the existence of much confusion which has arisen from the non-acceptance of the distinction made by Avogadro between atom and molecule ; we find the analogies between chemical affinity and electrical energy made the basis of a system of classification which regards every compound atom (or molecule) as built up of two parts, in one of which positive, and in the other negative electricity predominates; and accompanying this system of classification we find that an acid is no longer regarded as necessarily an oxygen compound, but rather as a compound possessed of certain properties which are probably due to the arrangement of the elementary atoms, among which hydrogen appears generally to find a place ; we find that salts are for the most part regarded as metallic derivatives of acids ; and we find that by the decomposition of the supposed elementary substances, potash, soda, lime, etc., the number of the elements has been extended, the application of a new instrument of research has been brilliantly rewarded, and the Lavoisierian description of " element " as the " attained, not the attainable, limit of research " has been emphasized.

CHAPTER V.

THE WORK OF GRAHAM.

Thomas Graham, 1805–1869.

THE work of Graham, concerned as it mostly was with the development of the conception of atoms, connects the time of Dalton with that in which we are now living. I have therefore judged it advisable to devote a short chapter to a consideration of the life-work of this chemist, before proceeding to the third period of chemical advance, that, namely, which witnessed the development of organic chemistry through the labours of men who were Graham's contemporaries.

The printed materials which exist for framing the story of Graham's life are very meagre, but as he appears, from the accounts of his friends, to have devoted himself entirely to scientific researches, we cannot go far wrong in regarding the history of his various discoveries as also the history of his life.

THOMAS GRAHAM was born in Glasgow, on

December 21, 1805. His father, James Graham, a successful manufacturer, was in a position to give his son a good education. After some years spent in the ordinary school training, Gràham entered Glasgow University at the early age of fourteen, and graduated as M.A. five years later. It was the intention of Graham's father that his son should enter the Scottish Church; but under the teaching of Dr. Thomas Thomson and others the lad imbibed so strong a love of natural science, that rather than relinquish the pursuit of his favourite study, he determined to be independent of his father and make a living for himself. His father was much annoyed at the determination of his son to pursue science, and vainly attempted to force him into the clerical profession. The quarrel between father and son increased in bitterness, and notwithstanding the intervention of friends the father refused to make his son any allowance for his maintenance; and although many years after a reconcilement was effected, yet at the time when Graham most needed his father's help he was left to struggle alone. Graham went to Edinburgh, where he pursued his studies under Hope and Leslie, professors of chemistry and physics respectively—men whose names were famous wherever natural science was studied. Graham's mother, for whom he had always the greatest respect and warmest love, and his sister Margaret helped him as best they could during this trying time.

The young student found some literary occupation and a little teaching in Edinburgh, and sometimes he was asked to make investigations in subjects connected with applied chemistry. Thus he struggled on for four or five years, during which time he began to publish papers on chemico-physical subjects. In the year 1829 he was appointed Lecturer on Chemistry at the Mechanics' Institution in Glasgow, and next year he was removed to the more important position of lecturer on the same science at the Andersonian Institution in that city. This position he occupied for seven years, when he was elected Professor of Chemistry in the University of London (now University College) : he had been elected to the Fellowship of the Royal Society in the preceding year. During his stay at the Andersonian Institution Graham had established his fame as a physical chemist ; he had begun his work on acids and salts, and had established the fundamental facts concerning gaseous diffusion. These researches he continued in London, and from 1837 to 1854 he enriched chemical science with a series of papers concerned for the most part with attempts to trace the movements of the atoms of matter.

In 1854 Graham succeeded Sir John Herschel in the important and honourable position of Master of the Mint. For some years after his appointment he was much engaged with the duties of his office, but about 1860 he again returned to his atomic studies, and in his papers on "Transpiration of Liquids" and on " Dialysis" he did much in the application of physical methods to solve chemical problems, and opened up new paths, by travelling on which his successors greatly advanced the limits of the science of chemistry. Graham was almost always at work ; his holidays were " few and far between." By the year 1868 or so his general health began to grow feeble ; in the autumn of 1869, during a visit to Malvern where he sought repose and invigorating air, he caught cold, which developed into inflammation of the lungs. On his return to London the disease was overcome by medical remedies, but he continued very weak, and gradually sank, till the end came on the 16th of September 1869.

I have said that the seven years during which Graham held the lectureship on chemistry in the Andersonian Institution, Glasgow, witnessed the beginning alike of his work on salts and of that on gaseous diffusion. He showed that there exists a series of compounds of various salts, *e.g.* chloride of calcium, chloride of zinc, etc., with alcohol. He compared the alcohol in these salts, which he called *alcoates*, to the water in ordinary crystallized salts, and thus drew the attention of chemists to the important part played by water in determining the properties of many substances. Three years later (1833) appeared one of his most important papers, bearing on the general conception of acids : " Researches on the Arseniates, Phosphates, and Modifications of Phosphoric Acid." Chemists at this time knew that phosphoric acid—that is, the substance obtained by adding water to pentoxide of phosphorus—exhibited many peculiarities, but they were for the most part content to leave these unexplained. Graham, following up the analogy which he had already established between water and bases, prepared and carefully determined the composition of a series of phosphates, and concluded that pentoxide of phosphorus is able to combine with a base—say soda—in three different proportions, and thus to produce three different phosphates of soda. But as Graham accepted that view which regards a salt as a metallic derivative of an acid, he supposed that three different phosphoric acids ought to exist ; these acids he found in the substances produced by the action of water on the oxide of phosphorus. He showed that just as the oxide combines with a base in three proportions, so does it combine with water in three proportions. This water he regarded as chemically analogous to the base in the three salts, one atom (we should now rather say molecule) of base could be replaced by one atom of water, two atoms of base by two atoms of water, or three atoms of base by three atoms of water. Phosphoric acid was therefore regarded by Graham as a compound of pentoxide of phosphorus and water, the latter being as essentially a part of the acid as the former. He distinguished between *monobasic, dibasic,* and *tribasic* phosphoric acids : by the action of a base on the *monobasic acid,* one, and only one salt was produced ; the *dibasic acid* could furnish two salts, containing different proportions (or a different number of atoms) of the same base : and from the *tribasic acid* three salts, containing the same base but in different proportions, could be obtained.

Davy's view of an acid as a compound of water with a negative oxide was thus confirmed, and there was added to chemical science the conception of *acids of different basicity.*

In 1836 Graham's paper on "Water as a Constituent of Salts" was published in the "Transactions of the Royal Society of Edinburgh." In this paper he inquires whether the water in crystalline salts can or cannot be removed without destroying the chemical individuality of the salts. He finds that in some crystalline salts part of the water can be easily removed by the application of heat, but the remainder only at very high temperatures. He distinguishes between those atoms of water which essentially belong to the compound atom of the salt, and those atoms which can be readily removed therefrom, which are as it were added on to, or built up around the exterior of the atom of salt. In this paper Graham began to distinguish what is now called *water of crystallization* from *water of constitution*, a distinction pointed to by some of Davy's researches, but a distinction which has remained too much a mere matter of nomenclature since the days of Graham.

In these researches Graham emphasized the necessity of the presence of hydrogen in all true acids ; as he had drawn an analogy between water and bases, so now he saw in the hydrogen of acids the analogue of the metal of salts. He regarded the structure of the compound atom of an acid as similar to that of the compound atom of a salt; the hydrogen atom, or atoms, in the acid was replaced by a metallic atom, or atoms, and so a compound atom of the salt was produced.

Davy and Berzelius had proved that hydrogen is markedly electro-positive; hydrogen appeared to Graham to belong to the class of metals. In making this bold hypothesis Graham necessarily paid little heed to those properties of metals which appeal to the senses of the observer. Metals, as a class, are lustrous, heavy, malleable substances; hydrogen is a colourless, inodourless, invisible, very light gas: how then can hydrogen be said to be metallic?

I have again and again insisted on the need of imagination for the successful study of natural science. Although in science we deal with phenomena which we wish to measure and weigh and record in definite and precise language, yet he only is the successful student of science who can penetrate beneath the surface of things, who can form mental pictures different from those which appear before his bodily eye, and so can discern the intricate and apparently irregular analogies which explain the phenomena he is set to study.

Graham was not as far as we can learn endowed, like Davy, with the sensitive nature of a poet, yet his work on hydrogen proves him to have possessed a large share of the gift of imagination. Picturing to himself the hydrogen atom as essentially similar in its chemical functions to the atom of a metal, he tracked this light invisible gas through many tortuous courses : he showed how it is absorbed and retained (*occluded* as he said) by many metals ; he found it in meteors which had come from far-away regions of space; and at last, the year before he died he prepared an alloy of palladium and the metal hydrogen, from which a few medals were struck, bearing the legend " Palladium-Hydrogenium 1869."

Within the last few years hydrogen has been liquified and, it is said, solidified. Solid hydrogen is described as a steel-grey substance which fell upon the table with a sound like the ring of a metal.

But Graham's most important work was concerned with the motion of the ultimate particles of bodies.

He uses the word "atom" pretty much as Dalton did. He does not make a distinction between the atom of an element and the atom of a compound, but apparently uses the term as a convenient one to express the smallest undivided particle of any chemical substance which exhibits the properties of that substance. As Graham was chiefly concerned with the physical properties of chemical substances, or with those properties which are studied alike by chemistry and physics, the distinction between atom and molecule, so all-important in pure chemistry, might be, and to a great extent was, overlooked by him. In considering his work we shall however do well to use the terms "atom" and "molecule" in the sense in which they are now always used in chemistry, a sense which has been already discussed (see pp. 139–143).

Many years before Graham began his work a curious fact had been recorded but not explained. In 1823 Döbereiner filled a glass jar with hydrogen and allowed the jar to stand over water: on returning after twelve hours he found that the water had risen about an inch and a half into the jar. Close examination of the jar showed the presence of a small crack in the glass. Many jars, tubes and flasks, all with small cracks in the glass, were filled with hydrogen and allowed to stand over water; in every case the water rose in the vessel. No rise of the water was however noticeable if the vessels were filled with ordinary air, nitrogen or oxygen.

In 1831 Graham began the investigation of the peculiar phenomenon observed by Döbereiner. Repeating Döbereiner's experiments, Graham found that a portion of the hydrogen in the cracked vessels passed outwards through the small fissures, and a little air passed inwards : the water therefore rose in the jar, tube or flask, because there was a greater pressure on the surface of the water outside than upon that inside the vessel. Any gas lighter than air behaved like hydrogen ; when gases heavier than air were employed the level of the water inside the vessel was slightly lowered after some hours.

Graham found that the passage of gases through minute openings could be much more accurately studied by placing the gas to be examined in a glass tube one end of which was closed by a plug of dry plaster of Paris, than by using vessels with small fissures· in the glass.

The *diffusion-tube* used by Graham generally consisted of a piece of glass tubing, graduated in fractions of a cubic inch and having a bulb blown near one end ; the short end was closed by a thin plug of dry plaster of Paris (gypsum), the tube was filled with the gas to be examined, and the open end was immediately immersed in water. The water was allowed to rise until it had attained a constant level, when it was found that the whole of the gas originally in the tube had passed outwards through the porous plug, and air had passed inwards. The volume of gas originally in the tube being known, and the volume of air in the tube at the close of the experiment being measured, it was only necessary to divide the former by the latter number in order to obtain the number of volumes of gas which had passed outwards for each one volume of air which had passed inwards ; in other words to obtain the *rate of diffusion* compared with air of the gas under examination.

Graham's results were gathered together in the statement, " The diffusion-rates of any two gases are inversely as the square roots of their densities."

III.   R

Thus, take oxygen and hydrogen: oxygen is sixteen times heavier than hydrogen, therefore hydrogen diffuses four times more rapidly than oxygen. Take hydrogen and air: the specific gravity of hydrogen is 0·0694, air being 1 ; the square root of 0·0694 is 0·2635, therefore hydrogen will diffuse more rapidly than air in the ratio of 0·2635 : 1.

In the years 1846–1849 Graham resumed this inquiry ; he now distinguished between *diffusion*, or the passage of gases through porous plates, and *transpiration*, or the passage of gases through capillary tubes. He showed that if a sufficiently large capillary tube be employed the rate of transpiration of a gas becomes constant, but that it is altogether different from the rate of diffusion of the same gas. He established the fact that there is a connection of some kind between the transpiration-rates and the chemical composition of gases, and in doing this he opened up a field of inquiry by cultivating which many important results have been gained within the last few years, and which is surely destined to yield more valuable fruit in the future.

Returning to the diffusion of gases, Graham, after nearly thirty years' more or less constant labour, begins to speculate a little on the causes of the phenomena he had so studiously and perseveringly been examining. In his paper on "The Molecular Mobility of Gases," read to the Royal Society in 1863, after describing a new diffusion-tube wherein thin plates of artificial graphite were used in place of plaster of Paris, Graham says, "The pores of artificial graphite appear to be really so minute that a gas *in mass* cannot penetrate the plate at all. It seems that molecules only can pass; and they may be supposed to pass wholly unimpeded by friction, for the smallest pores that can be imagined to exist in the graphite must be tunnels in magnitude to the ultimate atom of a gaseous body." He then shortly describes the molecular theory of matter, and shows how this theory—a sketch of which so far as it concerns us in this book has been given on pp. 123–125—explains the results which he has obtained. When a gas passed through a porous plate into a vacuum, or when one gas passed in one direction and another in the opposite direction through the same plate, Graham saw the molecules of each gas rushing through the "tunnels" of graphite or stucco. The average rate at which the molecules of a gas rushed along was the diffusion-rate of that gas. The lighter the gas the more rapid was the motion of its molecules. If a mixture of two gases, one much lighter than the other, were allowed to flow through a porous plate, the lighter gas would pass so much more quickly than the heavier gas that a partial separation of the two might probably be effected. Graham accomplished such a separation of oxygen and hydrogen, and of oxygen and nitrogen; and he described a simple instrument whereby this process of *atmolysis*, as he called it, might be effected.

Graham's *tube atmolyser* consisted of a long tobacco-pipe stem placed inside a rather shorter and considerably wider tube of glass; the pipe stem was fixed by passing through two corks, one at each end of the glass tube ; through one of these corks there also passed a short piece of glass tubing. When the instrument was employed, the piece of short glass tubing was connected with an air-pump, and one end of the pipe stem with the gaseous mixture—say ordinary air. The air-pump being set in motion, the gaseous mixture was allowed to flow slowly through the pipe stem ; the lighter ingredient of the mixture passed outwards through the pipe stem into the wide glass tube more rapidly than the heavier ingredient, and was swept away to the air-pump ; the heavier ingredient could be collected, mixed with only a small quantity of the lighter, at the other end of the pipe stem. As Graham most graphically expressed it, " The stream of gas diminishes as it proceeds, like a river flowing over a pervious bed."

Graham then contrived a very simple experiment whereby he was able to measure the rate of motion of the molecules of carbonic acid. He introduced a little carbonic acid into the lower part of a tall cylindrical jar, and at the close of certain fixed periods of time he determined the amount of carbonic acid which had diffused upwards through the air into the uppermost layer of the jar. Knowing the height of the jar, he now knew the distance through which a small portion of

carbonic acid passed in a stated time, and regarding this small portion as consisting of a great many
molecules, all moving at about equal rates, he had
determined the average velocity of the molecules
of carbonic acid. A similar experiment was performed with hydrogen. The general results were
that the molecules of carbonic acid move about
in still air with a velocity equal to seventy-three
millimetres per minute, and that under the same
conditions the molecules of hydrogen move with a
velocity equal to about one-third of a metre per
minute.*

The Bakerian Lecture for 1849, read by Graham
before the Royal Society, was entitled " On the
Diffusion of Liquids." In this paper he describes
a very large number of experiments made with a
view to determine the rate at which a salt in
aqueous solution diffuses, or passes upwards into
a layer of pure water above it, the salt solution
and the water not being separated by any inter-
vening medium. Graham's method of procedure
consisted in completely filling a small bottle with
a salt solution of known strength, placing this
bottle in a larger graduated vessel, and carefully
filling the latter with water. Measured portions of
the water in the larger vessel were withdrawn at
stated intervals, and the quantity of salt in each
portion was determined. Graham found that
under these conditions salts diffused with very

* A metre is equal to about thirty-nine inches ; a millimetre is the
one-thousandth part of a metre.

varying velocities. Groups of salts showed equal
rates of diffusion. There appeared to be no
definite connection between the molecular weights
of the salts and their diffusion-rates; but as
Graham constantly regarded diffusion, whether of
gases or liquids, as essentially due to the movements of minute particles, he thought that the
particles which moved about as wholes during
diffusion probably consisted of groups of what
might be called chemical molecules — in other
words, Graham recognized various orders of small
particles. As the atom was supposed to have a
simpler structure than the molecule (if indeed it
had a structure at all), so there probably existed
groups of molecules which, under certain conditions, behaved as individual particles with definite
properties.

As Graham applied the diffusion of gases to the
separation of two gases of unequal densities, so he
applied the diffusion of liquids to the separation
of various salts in solution. He showed also that
some complex salts, such as the alums, were partially separated into their constituents during the
process of diffusion.

The prosecution of these researches led to most
important results, which were gathered together in
a paper on "Liquid Diffusion applied to Analysis,"
read to the Royal Society in 1861.

Graham divided substances into those which diffused easily and quickly into water, and those
which diffused very slowly; he showed that the

former were all crystallizable substances, while the latter were non-crystallizable jelly-like bodies. Graham called these jelly-like substances *colloids ;* the easily diffusible substances he called *crystalloids.* He proved that a colloidal substance acts towards a crystalloid much as water does; that the crystalloid rapidly diffuses through the colloid, but that colloids are not themselves capable of diffusing through other colloids. On this fact was founded Graham's process of *dialysis.* As colloid he employed a sheet of parchment paper, which he stretched on a ring of wood or caoutchouc, and floated the apparatus so constructed—*the dialyser*—on the surface of pure water in a glass dish ; he then poured into the dialyser the mixture of substances which it was desired to separate. Let us suppose that this mixture contained sugar and gum ; the crystalloidal sugar soon passed through the parchment paper, and was found in the water outside, but the colloidal gum remained in the dialyser.

If the mixture in the dialyser contained two crystalloids, the greater part of the more diffusible of these passed through the parchment in a short time along with only a little of the less diffusible ; a partial separation was thus effected.

This method of dialysis was applied by Graham to separate and obtain in the pure state many colloidal modifications of chemical compounds, such as aluminium and tin hydrates, etc. By his study of these peculiar substances Graham introduced into chemistry a new class of bodies, and opened up great fields of research.

Matter in the colloidal state appears to be endowed with properties which are quite absent, or are hidden, when it is in the ordinary crystalloidal condition. Colloids are readily affected by the smallest changes in external conditions; they are eminently unstable bodies; they are, Graham said, always on the verge of an impending change, and minute disturbances in the surrounding conditions may precipitate this change at any moment. Crystalloids, on the other hand, are stable; they have definite properties, which are not changed without simultaneous large changes in surrounding conditions. But although, to use Graham's words, these classes of bodies "appear like different worlds of matter," there is yet no marked separating line between them. Ice is a substance which under ordinary conditions exhibits all the properties of crystalloids, but ice formed in contact with water just at the freezing point is not unlike a mass of partly dried gum; it shows no crystalline structure, but it may be rent and split like a lump of glue, and, like glue, the broken pieces may be pressed together again and caused to adhere into one mass.

"Can any facts," asks Graham, "more strikingly illustrate the maxim that in Nature there are no abrupt transitions, and that distinctions of class are never absolute?"

In the properties of colloids and crystalloids

Graham saw an index of diversity of molecular structure. The smallest individual particle of a colloid appeared to him to be a much more complex structure than the smallest particle of a crystalloid. The colloidal molecule appeared to be formed by the gathering together of several crystalloidal molecules; such a complex structure might be expected readily to undergo change, whereas the simpler molecule of a crystalloid would probably present more definite and less readily altered properties.

In this research Graham had again, as so often before, arrived at the conception of various orders of small particles. In the early days of the Daltonian theory it seemed that the recognition of atoms as ultimate particles, by the placing together of which masses of this or that kind of matter are produced, would suffice to explain all the facts of chemical combinations; but Dalton's application of the term "atom" to elements and compounds alike implied that an atom might itself have parts, and that one atom might be more complex than another. The way was thus already prepared for the recognition of more than one order of atoms, a recognition which was formulated three years after the appearance of Dalton's "New System" in the statement of Avogadro, "Equal volumes of gases contain equal numbers of molecules;" for we have seen that the application of this statement to actually occurring reactions between gases obliges us to admit that the molecules of hydrogen, oxygen

and many other elementary gases are composed
of two distinct parts or atoms.

Berzelius it is true did not formally accept the
generalization of Avogadro ; but we have seen how
the conception of atom which runs through his
work is not that of an indivisible particle, but
rather that of a little individual part of matter
with definite properties, from which the mass of
matter recognizable by our senses is constructed,
just as the wall is built up of individual bricks.
And as the bricks are themselves constructed of
clay, which in turn is composed of silica and
alumina, so may each of these little parts of matter
be constructed of smaller parts ; only as clay is not
brick, and neither silica nor alumina is clay, so the
properties of the parts of the atom—if it has parts
—are not the properties of the atom, and a mass
of matter constructed of these parts would not
have the same properties as a mass of matter constructed of the atoms themselves.

Another feature of Graham's work is found in
the prominence which he gives to that view of
a chemical compound which regards it as the
resultant of the action and reaction of the parts
of the compound. As the apparent stability of
chemical compounds was seen by Davy to be the
result of an equilibrium of contending forces, so
did the seemingly changeless character of any
chemical substance appear to Graham as due
to the orderly changes which are continually proceeding among the molecules of which the substance is constructed.

A piece of lime, or a drop of water, was to the mind of Graham the scene of a continual strife, for that minute portion of matter appeared to him to be constructed of almost innumerable myriads of little parts, each in more or less rapid motion, one now striking against another and now moving free for a little space. Interfere with those movements, alter the mutual action of those minute particles, and the whole building would fall to pieces.

For more than thirty years Graham was content to trace the movements of molecules. During that time he devoted himself, with an intense and single-minded devotion, to the study of molecular science. Undaunted in early youth by the withdrawal of his father's support; unseduced in his middle age by the temptations of technical chemistry, by yielding to which he would soon have secured a fortune; undazzled in his later days by the honours of the position to which he had attained; Graham dedicated his life to the nobler object of advancing the bounds of natural knowledge, and so adding to those truths which must ever remain for the good and furtherance of humanity.

CHAPTER VI.

RISE AND PROGRESS OF ORGANIC CHEMISTRY— PERIOD OF LIEBIG AND DUMAS.

Justus Liebig, 1803–1873. Jean Baptiste André Dumas, born in 1800.

I HAVE as yet said almost nothing with regard to the progress of organic chemistry, considered as a special branch of the science. It is however in this department that the greatest triumphs which mark the third period of chemical advance have been won. We must therefore now turn our attention to the work which has been done here.

The ancients drew no such distinction between portions of their chemical knowledge, limited as it was, as is implied by the modern terms " organic " and " inorganic chemistry." An organic acid—acetic —was one of the earliest known substances belonging to the class of acids ; many processes of chemical handicraft practised in the olden times dealt with the manufacture of substances, such as soap, leather or gum, which we should now call organic substances. Nor did the early alchemists, although working chiefly with mineral or inorganic substances, draw any strict division between the two branches of chemistry. The medical chemists of the sixteenth century dealt much with substances derived from plants and animals, such as benzoic and succinic acids, spirit of wine, oils, etc. But neither in their nomenclature nor in their practice did they sharply distinguish inorganic from organic compounds. They spoke of the *quintessence* of arsenic and the *quintessence* of alcohol ; they applied the term " oil " alike to the products of the action of acids on metallic salts and to substances obtained from vegetables. But towards the end of the seventeenth century, at the time that is when the phlogistic theory began to gain pre-eminence, we find gradually springing up a division of chemical substances into mineral, animal and vegetable substances—a division which was based rather on a consideration of the sources whence the substances were derived than on the properties of the substances themselves, and therefore a division which was essentially a non-chemical one.

About a century after this, systematic attempts began to be made to trace some peculiarity of composition as belonging to all compounds of organic, that is, of animal or vegetable, origin. As very many of the substances then known belonging to this class were more or less oil-like in their

properties—oils, fats, balsams, gums, sugar, etc.—
organic substances generally were said to be
characterized by the presence in them of the
principle of oil.

Such a statement as this, although suited to the
conceptions of that time, could not be received
when Lavoisier had shown chemists how Nature
ought be examined. With the definite conception
of element introduced by the new chemistry, came
an attempt to prove that organic compounds were
built up of elements which were rarely found
together in any one compound of inorganic origin.
Substances of vegetable origin were said by Lavoisier to be composed of carbon, hydrogen and
oxygen, while phosphorus and nitrogen, in addition
to those three elements, entered into the composition of substances derived from animals. But
neither could this definition of organic compounds
be upheld in the face of facts. Wax and many oils
contained only carbon and hydrogen, yet they
were undoubtedly substances of vegetable or animal
origin. If the presence of any two of the three
elements, carbon, hydrogen and oxygen, were to be
regarded as a sufficient criterion for the classification of a compound, then it was necessary that
carbonic acid—obtained by the action of a mineral
acid on chalk—should be called an organic compound.

To Berzelius belongs the honour of being the
chemist who first applied the general laws of
chemical combination to all compounds alike,

whether derived from minerals, animals, or vegetables. The ultimate particles, or molecules, of
every compound were regarded by Berzelius as
built up of two parts, each of which might itself be
an elementary atom, or a group of elementary
atoms. One of these parts, he said, was characterized by positive, the other by negative electricity.
Every compound molecule, whatever was the
nature or number of the elementary atoms composing it, was a dual structure (see p. 164).
Organic chemistry came again to be a term somewhat loosely applied to the compounds derived
from animals or vegetables, or in the formation
of which the agency of living things was necessary.
Most, if not all of these compounds contained
carbon and some other element or elements, especially hydrogen, oxygen and nitrogen.

But the progress of this branch of chemistry was
impeded by the want of any trustworthy methods
for analysing compounds containing carbon, oxygen
and hydrogen. This want was to be supplied, and
the science of organic chemistry, and so of chemistry
in general, was to be immensely advanced by the
labours of a new school of chemists, chief among
whom were Liebig and Dumas.

Let us shortly trace the work of these two
renowned naturalists. The life-work of the first is
finished ; I write this story of the progress of his
favourite science on the eighty-second birthday of
the second of these great men, who is still with us
a veteran crowned with glory, a true soldier in the

battle against ignorance and so against want and crime.

JUSTUS LIEBIG was born at Darmstadt, on the 12th of May 1803. The main facts which mark his life regarded apart from his work as a chemist are soon told. Showing a taste for making experiments he was apprenticed by his father to an apothecary. Fortunately for science he did not long remain as a concoctor of drugs, but, was allowed to enter the University of Bonn as a student of medicine. From Bonn he went to Erlangen, at which university he graduated in 1821. A year or two before this time Liebig had begun his career as an investigator of Nature, and he had already made such progress that the Grand Duke of Hesse-Darmstadt was prevailed on to grant him a small pension and allow him to prosecute his researches at Paris, which was then almost the only place where he could hope to find the conditions of success for the study of scientific chemistry. To Paris accordingly he went in 1823. He was so fortunate—thanks to the good graces of the renowned naturalist Alexander von Humboldt—as to be allowed to enter the laboratory of Gay-Lussac, where he continued the research on a class of explosive compounds, called *fulminates*, which he had begun before leaving Darmstadt.

A year later Liebig was invited to return to his native country as Professor of Chemistry in the small University of Giessen—a name soon to be known wherever chemistry was studied, and now held dear by many eminent chemists who there learned what is meant by the scientific study of Nature.

The year before Liebig entered the laboratory of Gay-Lussac there came to Paris a young and enthusiastic student who had already made himself known in the scientific world by his physiological researches, and who was now about to begin his career as a chemist.

In that southern part of France which is rich in memories of the Roman occupation, not far from the remains of the great aqueduct which spans the valley of the Gardon, at no great distance from the famous cities of Arles and Nîmes, was born, in the town of Alais, on the 14th of July 1800, JEAN BAPTISTE ANDRÉ DUMAS.

The father of Dumas was a man of considerable culture ; he gave his son as good an education as could be obtained in the little town of his birth. At the age of fourteen young Dumas was a good classical scholar, and had acquired a fair knowledge of natural science. But for his deficiency in mathematics he would probably have entered for the examination which admitted those who passed it to join the French navy. But before he had made good his mathematical deficiencies the troublous nature of the times (1814–15) obliged his parents to think of some other profession for their son which would entail less sacrifice on their part.

Like his great fellow-worker in after life he was

III.   S

apprenticed to an apothecary, and like him also, he soon forsook this sphere of usefulness.

Desirous of better opportunities for the study of science, and overpowered by the miseries which war had brought upon the district of his birth, Dumas persuaded his father to allow him to go to Geneva. At Geneva Dumas found an atmosphere more suited to his scientific progress; chemistry, physics, botany, and other branches of natural science were taught by men whose names were everywhere known. He began experiments in chemistry with the crudest and most limited apparatus, but even with these he made discoveries which afterwards led to important work on the volumes occupied by the atoms of elementary substances.

About the year 1818 Dumas became acquainted with Dr. J. L. Prévost, who had returned from studying in many of the most famous medical schools of Europe. Invited by Prévost to join in an investigation requiring medical, botanical and chemical knowledge, Dumas now began a series of researches which soon passed into the domain of animal physiology, and by the prosecution of which under many difficulties he laid the foundations of his future fame.

But along with his physiological work Dumas carried on a research into the expansion of various ethers. This necessitated the preparation of a series of ethers in a state of purity; but so difficult did Dumas find this to be, so much time did he consume in this preliminary work, and so interested did he become in the chemical part of the investigation, that he abandoned the experiments on expansion, and set himself to solve some of the problems presented by the composition and chemical properties of the ethers.

Dumas would probably have remained in Geneva had he not had a morning visit paid him in the year 1822. When at work in his laboratory one day, some one knocked and was bidden come in. " I was surprised to find myself face to face with a gentleman in a light-blue coat with metal buttons, a white waistcoat, nankeen breeches, and top-boots. . . . The wearer of this costume, his head somewhat bent, his eyes deep-set but keen, advanced with a pleasant smile, saying, ' Monsieur Dumas.' ' The same, sir ; but excuse me.' ' I am M. de Humboldt, and did not wish to pass through Geneva without having had the pleasure of seeing you.' . . . I had only one chair. My visitor was pleased to accept it, whilst I resumed my elevated perch on the drawing stool. . . . ' I intend,' said M. de Humboldt, ' to spend some days in Geneva, to see old friends and to make new ones, and more especially to become acquainted with young people who are beginning their career. Will you act as my cicerone ? I warn you however that my rambles begin early and end late. Now, could you be at my disposal, say from six in the morning till midnight ? ' " After some days spent as Humboldt had indicated the great naturalist left Geneva.

Dumas tells us that the town seemed empty to him. "I felt as if spell-bound. The memorable hours I had spent with that irresistible enchanter had opened a new world to my mind." Dumas felt that he must go to Paris—that there he would have more scope and more opportunities for prosecuting science. A few kind words, a little genuine sympathy, and a little help from Humboldt were thus the means of fairly launching in their career of scientific inquiry these two young men, Liebig and Dumas.

In Paris, whither he went in 1823, Dumas found a welcome. He soon made the acquaintance and gained the friendship of the great men who then made natural science so much esteemed in the French capital. When the year 1826 came, it saw him Professor of Chemistry at the Athenæum, and married to the lady whom he loved, and who has ever since fought the battle of life by his side.

Liebig left Paris in 1824. By the year 1830 he had perfected and applied that method for the analysis of organic compounds which is now in constant use wherever organic chemistry is studied ; by the same year Dumas had given the first warning of the attack which he was about to make on the great structure of dualism raised by Berzelius. In a paper, "On Some Points of the Atomic Theory," published in 1826, Dumas adopted the distinction made by Avogadro between molecules and atoms, or between the small particles of substances which remain undivided during physical actions, and the particles, smaller than these, which are undivided during chemical actions. But, unfortunately, Dumas did not mark these two conceptions by names sufficiently definite to enable the readers of his memoir to bear the distinction clearly in mind. The terms "atom" and "molecule" were not introduced into chemistry with the precise meanings now attached to them until some time after 1826.

Although the idea of two orders of small particles underlies all the experimental work described by Dumas in this paper, yet the numbers which he obtained as representing the actual atomic weights of several elements—*e.g.* phosphorus, arsenic, tin, silicon—show that he had not himself carried out Avogadro's hypothesis to its legitimate conclusions.

Two years after this Dumas employed the reaction wherein two volumes of gaseous hydrochloric acid are produced by the union of one volume of hydrogen with one volume of chlorine, as an argument which obliged him to conclude that, if Avogadro's physical hypothesis be accepted, the molecules of hydrogen and chlorine split, each into two parts, when these gases combine chemically. But Dumas did not at this time conclude that the molecular weight of hydrogen must be taken as twice its atomic weight, and that—hydrogen being the standard substance—the molecular weights of all gases must be represented by the specific gravities of these gases, referred to hydrogen as 2.

I have already shortly discussed the method for finding the relative weights of elementary atoms

which is founded on Avogadro's hypothesis, and, I
think, have shown that this hypothesis leads to the
definition of "atom" as the smallest amount of an
element in one molecule of any compound of that
element (see p. 142).

This deduction from Avogadro's law is now
a part and parcel of our general chemical knowledge. We wonder why it was not made by Dumas ;
but we must remember that a great mass of facts
has been accumulated since 1826, and that this
definition of "atom" has been gradually forced on
chemists by the cumulative evidence of those
facts.

One thing Dumas did do, for which the thanks
of every chemist ought to be given him ; he saw
the need of a convenient method for determining
the densities of compounds in the gaseous state, and
he supplied this need by that simple, elegant and
trustworthy method, still in constant use, known as
Dumas's vapour density process.

While Dumas was working out the details of
this analytical method, which was destined to be so
powerful an instrument of research, Liebig was
engaged in similar work ; he was perfecting that
process for the analysis of organic compounds
which has since played so important a part in
the advancement of this branch of chemical science.
The processes in use during the first quarter of
this century for determining the amounts of carbon,
hydrogen, and oxygen in compounds of those
elements, were difficult to conduct and gave un-

trustworthy results. Liebig adopted the principle
of the method used by Lavoisier, viz. that the
carbon in a compound can be oxidized, or burnt,
to carbonic͘ acid, and the hydrogen to water. He
contrived a very simple apparatus wherein this
burning might be effected and the products of
the burning—carbonic acid and water—might be
arrested and weighed. Liebig's apparatus remains
now essentially as it was presented to the chemical
world in 1830. Various improvements in details
have been made; the introduction of gas in place
of charcoal as a laboratory fuel has given the
chemist a great command over the process of
combustion, but in every part of the apparatus
to-day made use of in the laboratory is to be
traced the impress of the master's hand. A
weighed quantity of the substance to be analyzed
is heated with oxide of copper in a tube of
hard glass; the carbon is burnt to carbonic acid
and the hydrogen to water at the expense of the
oxygen of the copper oxide. Attached to the combustion tube is a weighed tube containing chloride
of calcium, a substance which greedily combines
with water, and this tube is succeeded by a set of
three or more small bulbs, blown in one piece of
glass, and containing an aqueous solution of caustic
potash, a substance with which carbonic acid
readily enters into combination. The chloride of
calcium tube and the potash bulbs are weighed
before and after the experiment; the increase in
weight of the former represents the amount of

water, and the increase in weight of the latter the amount of carbonic acid obtained by burning a given weight of the compound under examination. As the composition of carbonic acid and of water is known, the amounts of carbon and of hydrogen in one hundred parts of the compound are easily found; the difference between the sum of these and one hundred represents the amount of oxygen in one hundred parts of the compound. If the compound should contain elements other than these three, those other elements are determined by special processes, the oxygen being always found by difference.

Soon after his settlement at Giessen Liebig turned his attention to a class of organic compounds known as the *cyanates;* but Wöhler—who, while Liebig was in Paris in the laboratory of Gay-Lussac, was engaged in studying the intricacies of mineral chemistry under the guidance of Berzelius—had already entered on this field of research. The two young chemists compared notes, recognized each other's powers, and became friends; this friendship strengthened as life advanced, and some of the most important papers which enriched chemical science during the next thirty years bore the joint signatures of Liebig and Wöhler.

I have already mentioned that when it was found necessary to abandon the Lavoisierian definition of organic chemistry as the chemistry of compounds containing carbon, hydrogen and oxygen, and sometimes also phosphorus or nitrogen, a definition was attempted to be based on the supposed fact that the formation of the compounds obtained from animals and plants could be accomplished only by the agency of a living organism. But the discovery made in 1828 by Wöhler, that *urea*— a substance specially characterized by its production in the animal economy, and in that economy only—could be built up from mineral materials, rendered this definition of organic chemistry impossible, and broke down the artificial barrier whereby naturalists attempted to separate two fields of study between which Nature made no division.

We have here another illustration of the truth of the conception which underlies so many of the recent advances of science, which is the central thought of the noble structure reared by the greatest naturalist of our time, and which is expressed by one of the profoundest students of Nature that this age has seen in the words I have already quoted from the preface to the " Lyrical Ballads," " In Nature everything is distinct, but nothing defined into absolute independent singleness."

From this time the progress of organic chemistry became rapid. Dumas continued the researches upon ethers which he had commenced at Geneva, and by the year 1829 or so he had established the relations which exist between ethers and alcohols on the one hand, and ethers and acids on the other. This research, a description of the details of which

I cannot introduce here as it would involve the use of many technical terms and assume the possession by the reader of much technical knowledge, was followed by others, whereby Dumas established the existence of a series of compounds all possessed of the chemical properties of alcohol, all containing carbon, hydrogen and oxygen, but differing from one another by a constant amount of carbon and hydrogen. This discovery of a series of alcohols, distinguished by the possession of certain definite properties whereby they were marked off from all other so-called organic compounds, was as the appearance of a landmark to the traveller in a country where he is without a guide. The introduction of the comparative method of study into organic chemistry—the method, that is, which bases classification on a comparison of large groups of compounds, and which seeks to gather together those substances which are like and to separate those which are unlike—soon began to bear fruit. This method suggested to the experimenter new points of view from which to regard groups of bodies; analogies which were hidden when a few substances only were considered, became prominent as the range of view was widened. What the gentle Elia calls "fragments and scattered pieces of truth," "hints and glimpses, germs, and crude essays at a system," became important. There was work to be done, not only by the master spirits who, looking at things from a central position of vantage, saw the relative importance of the various detailed facts, but also by those who could only "beat up a little game peradventure, and leave it to knottier heads, more robust constitutions, to run it down."

Twenty years before the time of which we are now speaking Davy had decomposed the alkalis potash and soda ; as he found these substances to be metallic oxides, he thought it very probable that the other well-known alkali, ammonia, would also turn out to be the oxide of a metal. By the electrolysis of salts formed by the action of ammonia on acids, using mercury as one of the poles of the battery, Davy obtained a strange-looking spongy substance which he was inclined to regard as an alloy of the metallic base of ammonia with mercury. From the results of experiments by himself and others, Davy adopted a view of this alloy which regarded it as containing a *compound radicle*, or group of elementary atoms which in certain definite chemical changes behaved like a single elementary atom.

To this compound radicle he gave the name of *ammonium*.

As an aqueous solution of potash or soda was regarded as a compound of water and oxide of potassium or sodium, so an aqueous solution of ammonia was regarded as a compound of water and oxide of ammonium.

When the composition of this substance, ammonium, came to be more accurately determined, it was found that it might be best represented as a compound atom built up of one atom of nitrogen and four atoms of hydrogen. The observed properties of many compounds obtained from ammonia, and the analogies observed between these and similar compounds obtained from potash and soda, could be explained by assuming in the compound atom (or better, in the molecule) of the ammonia salt, the existence of this group of atoms, acting as one atom, called ammonium.

The reader will not fail to observe how essentially atomic is this conception of compound radicle. The ultimate particle, the molecule, of a compound has now come to be regarded as a structure built up of parts called atoms, just as a house is a structure built up of parts called stones and bricks, mortar and wood, etc. But there may be a closer relationship between some of the atoms in this molecule than between the other atoms. It may be possible to remove a group of atoms, and put another group—or perhaps another single atom— in the place of the group removed, without causing the whole atomic structure to fall to pieces ; just as it may be possible to remove some of the bricks from the wall of a house, or a large wooden beam from beneath the lintels, and replace these by other bricks or by a single stone, or replace the large wooden beam by a smaller iron one, without involving the downfall of the entire house. The group of atoms thus removable—the compound radicle—may exist in a series of compounds. As we have an oxide, a sulphide, a chloride, a nitrate, etc., of sodium, so we may have an oxide, a sulphide, a chloride, a nitrate, etc., of ammonium. The compounds of sodium are possessed of many properties in common ; this is partly explained by saying that they all contain one or more atoms of the element sodium. The compounds of ammonium possess many properties in common, and this is partly explained if we assume that they all contain one or more atoms of the compound radicle ammonium.

The conception of compound radicle was carried by Berzelius to its utmost limits. We have learned that the Swedish chemist regarded every molecule as composed of two parts ; in very many cases each of these parts was itself made up of more than one kind of atom—it was a compound radicle. But the Berzelian system tended to become too artificial : it drifted further and further away from facts. Of the two parts composing the dual molecular structure, one was of necessity positively, and the other negatively electrified. The greater number of the so-called organic compounds contained oxygen ; oxygen was the most electro-negative element known ; hence most organic compounds were regarded as formed by the coming together of one, two, or more atoms of oxygen, forming the negative part of the molecule, with one, two, or more atoms of a compound radicle, which formed the positive part of the molecule.

From this dualistic view of the molecule there naturally arose a disposition to regard the compound radicles of organic chemistry as the nonoxygenated parts of the molecules of organic compounds. An organic compound came gradually to be regarded as a compound of oxygen with some other elements, which were all lumped together under the name of a compound radicle, and organic chemistry was for a time defined as the chemistry of compound radicles.

From what has been said on p. 268, I think it will be evident that the idea of *substitution* is a necessary part of the original conception of compound radicle; a group of atoms in a molecule may, it is said, be removed, and another group, or another atom, *substituted* for that which is removed. Berzelius adopted this idea, but he made it too rigid; he taught that an electro-negative atom, or compound radicle, could be replaced or substituted only by another electro-negative atom or group of atoms, and a positively electrified atom or group of atoms, only by another electro-positive atom or compound radicle. Thus oxygen could perhaps be replaced by chlorine, but certainly not by hydrogen; while hydrogen might be replaced by a positively electrified atom, but certainly not by chlorine.

The conceptions of compound radicles and of substitution held some such position in organic chemistry as that which I have now attempted to indicate when Dumas and Liebig began their work in this field.

The visitors at one of the royal *soirées* at the Tuileries were much annoyed by the irritating vapours which came from the wax candles used

to illuminate the apartments; Dumas was asked
to examine the candles and find the reason of their
peculiar behaviour. He found that the manufacturer had used chlorine to bleach the wax, that
some of this chlorine remained in the candles, and
that the irritating vapours which had annoyed the
guests of Charles X. contained hydrochloric acid,
produced by the union of chlorine with part of the
hydrogen of the wax. Candles bleached by some
other means than chlorine were in future used in
the royal palaces; and the unitary theory, which
was to overthrow the dualism of Berzelius, began
to arise in the mind of Dumas.

The retention of a large quantity of chlorine by
wax could scarcely be explained by assuming that
the chlorine was present only as a mechanically
held impurity. Dumas thoroughly investigated
the action of chlorine on wax and other organic
compounds; and in 1834 he announced that
hydrogen in organic compounds can be exchanged
for chlorine, every volume of hydrogen given up by
the original compound being replaced by an equal
volume of chlorine.

Liebig and Wöhler made use of a similar conception to explain the results which they had
obtained about this time in their study of the oil
of bitter almonds, a study which will be referred to
immediately.

The progress of this bold innovation made by
Dumas was much advanced by the experiments
and reasonings of two French chemists, whose

names ought always to be reverenced by students of chemistry as the names of a pair of brilliant naturalists to whom modern chemistry owes much. *Gerhardt* was distinguished by clearness of vision and expression ; *Laurent* by originality, breadth of mind and power of speculation.

Laurent appears to have been the first who made a clear statement of the fundamental conception of the unitary theory: "Many organic compounds, when treated with chlorine lose a certain number of equivalents of hydrogen, which passes off as hydrochloric acid. An equal number of equivalents of chlorine takes the place of the hydrogen so eliminated ; thus the physical and chemical properties of the original substance are not profoundly changed. The chlorine occupies the place left vacant by the hydrogen ; the chlorine plays in the new compound the same part as was played by the hydrogen in the original compound."

The replacement of electro-positive hydrogen by electro-negative chlorine was against every canon of the dualistic chemistry ; and to say that the physical and chemical properties of the original compound were not profoundly modified by this replacement, seemed to be to call in question the validity of the whole structure raised by the labours during a quarter of a century of one universally admitted to be among the foremost chemists of his age.

But facts accumulated. By the action of

chlorine on alcohol Liebig obtained *chloroform* and *chloral*, substances which have since been so largely applied to the alleviation of human suffering; but it was Dumas who correctly determined the composition of these two compounds, and showed how they are related to alcohol and to one another.

Liebig's reception of the corrections made by Dumas in his work furnishes a striking example of the true scientific spirit. "As an excellent illustration," said Liebig, "of the mode in which errors should be corrected, the investigation of chloral by Dumas may fitly be introduced. It carried conviction to myself, as I think to everybody else, not by the copious number of analytical data opposed to the not less numerous results which I had published, but because these data gave a simpler explanation both of the formation and of the changes of the substances in question."

One of the most important contributions to the new views was made by Dumas in his paper on the action of chlorine on acetic acid (1833), wherein he proved that the product of this action, viz. *trichloracetic acid*, is related to the parent substance by containing three atoms of chlorine in place of three atoms of hydrogen in the molecule; that the new substance is, like the parent substance, a monobasic acid; that its salts are very analogous in properties to the salts of acetic acid; that the action of the same reagents on the two substances is similar; and finally, that the existence of many derivatives

III.   T

of these compounds could be foretold by the help
of the new hypothesis, which derivatives ought not
to exist according to the dualistic theory, but which,
unfortunately for that theory, were prepared and
analyzed by Dumas.

I have alluded to a research by Liebig and
Wöhler on oil of bitter almonds as marking an
important stage in the advance of the anti-dualistic
views. The paper alluded to was published in
1832. At that time it was known that *benzoic acid*
is formed by exposure of bitter-almond oil to the
air. Liebig and Wöhler made many analyses of
these two substances, and many experiments on the
mutual relations of their properties, whereby they
were led to regard the molecules of the oil as built up
each of an atom of hydrogen and an atom of a compound radicle—itself a compound of carbon, hydrogen and oxygen—to which they gave the name of
*benzoyl.** Benzoic acid they regarded as a compound of the same radicle with another radicle,
consisting of equal numbers of oxygen and hydrogen atoms. By the action of chlorine and other
reagents on bitter-almond oil these chemists obtained substances which were carefully analyzed
and studied, and the properties of which they

---

* "In reviewing once more the facts elicited by our inquiry, we
find them arranged around a common centre, a group of atoms pre-
serving intact its nature, amid the most varied associations with
other elements. This stability, this analogy, pervading all the phenomena, has induced us to consider this group as a sort of compound
element, and to designate it by the special name of *benzoyl*."—
Liebig and Wöhler, 1832.

showed could be simply explained by regarding them all as compounds of the radicle *benzoyl* with chlorine and other atoms or groups of atoms. But this view, if adopted, necessitated the belief that chlorine atoms could replace oxygen atoms; and, generally, that the substitution of an electro-positive by a negative atom or group of atoms did not necessarily cause any great alteration in the properties of the molecule.

Thus it was that the rigid conceptions of dualism were shown to be too rigid; that the possibility of an electro-positive radicle, or atom, replacing another of opposite electricity was recognized; and thus the view which regarded a compound molecule as one structure—atoms in which might be replaced by other atoms irrespective of the mutual electrical relations of these atoms—began to gain ground.

From this time the molecule of a compound has been generally regarded as a unitary structure, as one whole, and the properties of the molecule as determined by the nature, number, and arrangement of all the atoms which together compose it.

The unitary conception of a compound molecule appeared at first to be altogether opposed to the system of Berzelius; but as time went on, and as fresh facts came to be known, it was seen that the new view conserved at least one, and that perhaps the most important, of the thoughts which formed the basis of the Berzelian classification.

Underlying the dualism of Berzelius was the con-

ception of the molecule as an atomic structure;
this was retained in the unitary system of Dumas,
Gerhardt and Laurent.

Berzelius had insisted that every molecule is a
dual structure. This is taking too narrow a view
of the possibilities of Nature, said the upholders of
the new school. *This* molecule may have a dual
structure; *that* may be built up of three parts. The
structure of this molecule or of that can be determined only by a careful study of its relations with
other molecules.

For a time it seemed also as if the new chemistry
could do without the compound radicle which had
been so much used by Berzelius ; but the pressure
of facts soon drove the unitary chemists to recognize the value of that hypothesis which looked on
parts of the molecule as sometimes more closely
associated than other parts—which recognized the
existence of atomic structures within the larger
molecular structures. As a house is not simply
a putting together of so many bricks, so much
mortar, so many doors and windows, so many
leaden pipes, etc., but rather a definite structure
composed of parts, many of which are themselves
also definite structures, such as the window and its
accessory parts, the door with its lintel and handle,
etc., so to the unitary chemists did the molecule
appear to be built up of parts, some of which,
themselves composed of yet smaller parts, dis-
charged a particular function in the molecular
economy.

A general division of a plant might describe it as a structure consisting of a stem, a root, and leaves. Each of the parts, directly by its individual action and indirectly by the mutual action between it and all the other parts, contributes to the growth of the whole plant; but if the stem, or root, or leaves be further analyzed, each is found to consist of many parts, of fibres and cells and tissue, etc. We may liken the plant to the molecule of an organic compound; the root, the stem and the leaves to the compound radicles of which this molecule is built up, and the tissue, fibres, etc., to the elementary atoms which compose these compound radicles. The molecule is one whole, possessed of definite structure and performing a definite function by virtue of the nature and the arrangement of its parts.

Many years elapsed after the publication of the researches of Dumas, and of Liebig and Wöhler, before such a conception of the molecule as this was widely accepted by chemists. The opposition of the older school, headed by their doughty champion Berzelius, had to be overcome; the infallibility of some of the younger members of the new school had to be checked; facts had to be accumulated, difficulties explained, weak analogies abandoned and strong ones rendered stronger by research; special views of the structure of this or that molecule, deduced from a single investigation, had to be supplemented and modified by wider views gained by the researches of many workers. It was not

till 1867 that Liebig, when asked by Dumas at a
dinner given during the French Exhibition to the
foreign chemists, why he had abandoned organic
chemistry, replied that "now, with the theory of
substitution as a foundation, the edifice may be
built up by workmen: masters are no longer
needed."

Laurent and Gerhardt did noble work in advancing
the unitary theory; to them is largely due the
fruitful conception of types, an outcome of Dumas's
work, which owed its origin to the flickering of the
wax candles in the Tuileries during the royal *soirée*.

Chlorine can be substituted for hydrogen in acetic
acid, and the product is closely related in its properties to the parent substance; various atoms or
groups of atoms can be substituted by other groups
in the derivatives of oil of bitter almonds, but a
close analogy in properties runs through all these
compounds: these facts might be more shortly
expressed by saying that acetic and trichloracetic
acids belong to the same *type*, and that the derivatives of bitter-almond oil likewise belong to one
type.

Laurent carried this conception into inorganic
chemistry. Water and potash did not seem to have
much in common, but Laurent said potash is not a
compound of oxide of potassium and water, it is
rather a derivative of water. The molecule of potash
is derived from that of water by replacing one atom
of hydrogen in the latter by one atom of potassium;
water and potash belong to the same type.

Thus there was constituted *the water type.*

Light was at once thrown on many facts in organic chemistry. The analogies between alcohol and water, some of which were first pointed out by Graham (see p. 235), seemed to follow as a necessary consequence when the molecule of alcohol was regarded as built on the water type. In place of two atoms of hydrogen combined with one of oxygen, there was in the alcohol molecule one atom of the compound radicle *ethyl* (itself composed of carbon and hydrogen), one atom of oxygen and one of hydrogen. Alcohol was water with one hydrogen atom substituted by one ethyl atom ; the hydrogen atom was the atom of what we call an element, the ethyl was the atom of what we call a compound radicle.

Gerhardt sought to refer all organic compounds to one or other of three types—the water type, the hydrochloric acid type, and the ammonia type. As new compounds were prepared and examined, other types had to be introduced. To follow the history of this conception would lead us into too many details ; suffice it to say that the theory of types was gradually merged in the wider theory of equivalency, about which I shall have a little to say in the next chapter.

One result of the introduction of types into chemical science, associated as it was with the unitary view of compound radicles, was to overthrow that definition of organic chemistry which had for some time prevailed, and which stated that organic chemistry is "the chemistry of compound radicles." Compound radicles, it is true, were more used in explaining the composition and properties of substances obtained from animals and vegetables than of mineral substances, but a definition of one branch of a science which practically included the other branch, from which the first was to be defined, could not be retained. Chemists became gradually convinced that a definition of organic chemistry was not required; that there was no distinction between so-called organic and inorganic compounds; and they have consented, but I scarcely think will much longer consent, to retain the terms "organic" and "inorganic," only because these terms have been so long in use. The known compounds of the element carbon are so numerous, and they have been so much studied and so well classified, that it has become more convenient for the student of chemistry to consider them as a group, to a great extent apart from the compounds of the other elements; to this group he still often gives the name of "organic compounds."

Liebig continued to hold the chair of Chemistry in the University of Giessen until the year 1852, when he was induced by the King of Bavaria to accept the professorship of the same science in the University of Munich. During the second quarter of this century Giessen was much resorted to by students of chemistry from all parts of the world, more especially from England. Many men who

afterwards made their mark in chemical discovery
worked under the guidance of the professor of
Stockholm, but Giessen has the honour of being
the place where a well-appointed chemical laboratory for scientific research was first started as a
distinctly educational institution. The fame of
Liebig as a discoverer and as a teacher soon filled
the new institution with students, who were stirred
to enthusiasm as they listened to his lectures, or
saw him at work in his laboratory. "Liebig was
not exactly what is called a fluent speaker," says
Professor Hofmann, of Berlin, "but there was an
earnestness, an enthusiasm in all he said, which
irresistibly carried away the hearer. Nor was it so
much the actual knowledge he imparted which
produced this effect, as the wonderful manner in
which he called forth the reflective powers of even
the least gifted of his pupils. And what a boon
was it, after having been stifled by an oppressive
load of facts, to drink the pure breath of science
such as it flowed from Liebig's lips ! what a delight,
after having perhaps received from others a sack full
of dry leaves, suddenly in Liebig's lectures to see
the living, growing tree ! . . . We felt then, we feel
still, and never while we live shall we forget, Liebig's
marvellous influence over us ; and if anything could
be more astonishing than the amount of work he
did with his own hands, it was probably the mountain of chemical toil which he got us to go through.
Each word of his carried instruction, every intona-
tion of his voice bespoke regard ; his approval was

a mark of honour, and of whatever else we might be proud, our greatest pride of all was having him for our master. , . . Of our young winnings in the noble playground of philosophical honour, more than half were free gifts to us from Liebig, and to his generous nature no triumphs of his own brought more sincere delight than that which he took in seeing his pupils' success, and in assisting, while he watched, their upward struggle."

Liebig had many friends in England. He frequently visited this country, and was present at several meetings of the British Association. At the meeting of 1837 he was asked to draw up a report on the progress of organic chemistry ; he complied, and in 1840 presented the world with a book which marks a distinct epoch in the applications of science to industrial pursuits—"Chemistry in its Applications to Agriculture and Physiology."

In this book, and in his subsequent researches and works,* Liebig established and enforced the necessity which exists for returning to the soil the nourishing materials which are taken from it by the growth of crops ; he suggested that manure rich in the salts which are needed by plants might be artificially manufactured, and by doing this he laid the foundation of a vast industry which has arisen during the last two decades. He strongly and successfully attacked the conception which prevailed

* "Animal Chemistry, or Chemistry in its Applications to Physiology and Pathology," 1842. "Researches on the Chemistry of Food," 1847. " The Natural Laws of Husbandry," 1862.

among most students of physiology at that time, that chemical and physical generalizations could not be applied to explain the phenomena presented by the growth of living organisms. He was among the first to establish, as an induction from the results of many and varied experiments, the canon which has since guided all teachers of the science of life, that a true knowledge of biology must be based on a knowledge of chemistry and physics.

But Liebig was not content to establish broad generalizations and to leave the working out of them to others ; he descended from the heights of philosophical inquiry, and taught the housewife to make soup wherein the greatest amount of nourishment was conveyed to the invalid in the most easily digestible form ; and has he not, by bringing within the reach of every one a portion of the animal nourishment which else had run to waste in the pampas of South America or the sheep-runs of Australia, made his name, in every English home, familiar as a household word ?

On the death of Berzelius in 1848, it was to Liebig that every chemist looked for a continuation of the annual Report on the progress of chemistry, which had now become the central magazine of facts, whither each worker in the science could resort to make himself acquainted with what had been done by others on any subject which he proposed to investigate. From that time to the present day Liebig's *Annalen* has been the leading chemical journal of the world.

Of the other literary work of Liebig—of his essays, his celebrated " Chemical Letters," his many reports, his severe and sometimes harsh criticisms of the work of others—of the details of the three hundred original papers wherein he embodied the results of his researches, I have not time, nor would this be the place, to speak.

Honoured by every scientific society of any note in the world, crowned with the highest reward which England and France can offer to the man of science who is not an Englishman or a Frenchman—the Copley Medal and the associateship of the Institute—honoured and respected by every student of science, loved by each of the band of ardent natures whom he had trained and sent forth to battle for the good of their race, and, best of all, working himself to the last in explaining the wonders of Nature, he " passed into the silent land " on the 18th of April 1873, leaving the memory of a life nobly devoted to the service of humanity, and the imperishable record of many truths added to the common stock of the race.

The life-work of Dumas, other than that which I have already sketched, is so manifold and so varied, that to do more than refer to one or two leading points would carry us far beyond the limits within which I have tried to keep throughout this book. In one of his earliest papers Dumas adopted the atomic theory as the corner-stone of his chemical system ; he was thus led to an experimental revision of the values generally accepted for the atomic weights of some of the elements. Among these revisions, that of the atomic weight of carbon holds a most important place, partly because of the excellency of the work, but more because of the other inquiries to which this work gave rise.

Dumas's experiments were summed up in the statement that the atom of carbon is twelve times heavier than the atom of hydrogen. The experimental methods and the calculations used in this determination involved a knowledge of the atomic weight of oxygen; in order accurately to determine the value to be assigned to this constant, Dumas, in conjunction with Boussingault, undertook a series of experiments on the synthesis of water, which forms one of the classical researches of chemistry, and wherein the number 16 was established as representing the atomic weight of oxygen. Stas, from experiments conducted at a later time with the utmost care and under conditions eminently fitted to gain accurate results, obtained the number 15·96, in place of 16, for the atomic weight of oxygen; but in a paper recently published by the veteran Dumas, a source of error is pointed out which Stas had overlooked in his experiments, and it is shown that this error would tend slightly to increase the number obtained by Stas.

As the values assigned to the atomic weights of the elements are the very fundamental data of chemistry, and as we are every day more clearly perceiving that the mutual relations between the properties of elements and compounds are closely connected with the relative weights of the elementary atoms, we can scarcely lay too much stress on such work as this done by Dumas and Stas. Not many years after the publication of Dalton's "New System," the hypothesis was suggested by Prout that the atomic weights of all the elements are represented by whole numbers— that of hydrogen being taken as unity—that the atom of each element is probably formed by the putting together of two, three, four, or more atoms of hydrogen, and that consequently there exists but a single elementary form of matter. Among the upholders of this hypothesis Dumas has held an important place. He modified the original statement of Prout, and suggested that all atomic weights are whole multiples of half of that of hydrogen (that is, are whole multiples of $\frac{1}{2}$). The experiments of Stas seemed to negative this view, but later work—more especially the important critical revision of the results obtained by all the most trustworthy workers, conducted by Professor Clarke of Cincinnati, and published by the Smithsonian Institution as part of their series of "Constants of Nature"—has shown that we are in no wise warranted by facts in rejecting Prout's hypothesis as modified by Dumas, but that the balance of evidence is at present rather in its favour.

It would be altogether out of place to discuss

here an hypothesis which leads to some of the most abstruse speculations as to the nature of matter in which chemists have as yet ventured to indulge. I mention it only because it illustrates the far-reaching nature of the researches of the chemist whose work we are now considering, and also because it shows the shallowness of the scoffs in which some partly educated people indulge when they see scientific men occupying themselves for years with attempts to solve such a minute and, as they say, trivial question as whether the number 15·96 or the number 16 is to be preferred as representing the atomic weight of oxygen ; "for in every speck of dust that falls lie hid the laws of the universe, and there is not an hour that passes in which you do not hold the infinite in your hand."

Another and very different subject, which has been placed on a firm basis by the researches of Dumas, is the chemistry of fermentation. By his work on the action of beer-yeast on saccharine liquids, Dumas proved Liebig's view to be untenable—according to which the conversion of sugar into alcohol is brought about by the influence of chemical changes proceeding in the ferment ; also that the view of Berzelius, who regarded alcoholic fermentation as due simply to the contact of the ferment with the sugar, was opposed to many facts ; and lastly, Dumas showed that the facts were best explained by the view which regarded the change of sugar into alcohol as in no way different from other purely chemical changes, but as a change brought about, so far as our present knowledge goes, only by the agency of a growing organism of low form, such as yeast.

In 1832 Dumas established at his own expense a laboratory for chemical research. When the Revolution of 1848 broke out Dumas's means were much diminished, and he could no longer afford to maintain his laboratory. The closing of this place, where so much sound work had been done, was generally regarded as a calamity to science. About this time Dumas received a visit from a person of unprepossessing appearance, who accosted him thus: "They assert that you have shut up your laboratory, but you have no right to do so. If you are in need of money, there," throwing a roll of bank-notes on the table, "take what you want. Do not stint yourself; I am rich, a bachelor, and have but a short time to live." Dumas's visitor turned out to be Dr. Jecker. He assured Dumas that he was now only paying a debt, since he had made a fortune by what he had learnt in the medical schools of Paris. Dumas could not however in those troublous times turn his mind continuously to experimental research, and therefore declined Dr. Jecker's offer with many protestations of good will and esteem.

New work now began to press upon Dumas; his energy and his administrative powers were demanded by the State. Elected a member of the National Assembly in 1848, he was soon called by the President of the Republic to office as Minister of Agriculture and Commerce. He was made a senator under the second empire. He entered the municipal council of Paris about 1854, and was soon elected to the presidency. Under his presidency the great scheme for providing Paris with spring-water carried by aqueducts and tunnels was successfully accomplished ; many improvements were made in the drainage of the city ; the cost of gas was decreased, while the quality was improved, the constancy of the supply insured, and the appliances for burning the gas in the streets were altered and rendered more effective.

Nominated to succeed Pelouze as Master of the Mint in 1868, Dumas held this honourable and important position only until the Franco-German war of 1870. Since that date he has relinquished political life ; but as Permanent Secretary of the Academy Dumas now fills the foremost place in all affairs connected with science, whether pure or applied, in the French capital.

In the work of these two chemists, Liebig and Dumas, we find admirable illustrations of the scientific method of examining natural appearances.

In the broad general views which they both take of the phenomena to be studied, and the patient and persevering working out of details, we have shown us the combination of powers which are generally found in separate individuals.

Dumas has always insisted on the need of comparing properties and reactions of groups of bodies,

III.   U

before any just knowledge can be gained as to the position of a single substance in the series studied by the chemist. It has been his aim as a teacher, we are assured by his friend, Professor Hofmann, never to present to his students "an isolated phenomenon, or a notion not logically linked with others." To him each chemical compound is one in a series which connects it directly with many other similar compounds, and indirectly with other more or less dissimilar compounds.

Amid the overwhelming mass of facts which threaten nowadays to bury the science of chemistry, and crush the life out of it by their weight, Dumas tracks his way by the aid of general principles; but these principles are themselves generalized from the facts, and are not the offspring of his own fancy.

We have, I think, found that throughout the progress of chemical science two dangers have beset the student. He has been often tempted to accumulate facts, to amass analytical details, to forget that he is a chemist in his desire to perfect the instrument of analysis by the use of which he raises the scaffolding of his science; on the other hand, he has been sometimes allured from the path of experiment by his own day-dreams. The discoveries of science have been so wonderful, and the conceptions of some of those who have successfully prosecuted science have been so grand, that the student has not unfrequently been tempted to rest in the prevailing theories of the day, and, forgetting that these ought only "to afford peaceful lodgings to the intellect for the time," he has rather allowed them to circumscribe it, until at last the mind "finds difficulty in breaking down the walls of what has become its prison, instead of its home."

We may think that Dumas fell perhaps slightly into the former of these errors, when he did not allow his imagination a little more scope in dealing with the conception of "atom" and "molecule," the difference between which he had apprehended but not sufficiently marked by the year 1826 (see p. 261).

We know, from his own testimony, that Liebig once fell into the latter error and that the consequences were disastrous. "I know a chemist"— meaning himself—"who . . . undertook an investigation of the liquor from the salt-works. He found iodine in it, and observed, moreover, that the iodide of starch turned a fiery yellow by standing overnight. The phenomenon struck him ; he saturated a large quantity of the liquor with chlorine, and obtained from this, by distillation, a considerable quantity of a liquid which coloured starch yellow, and externally resembled chloride of iodine, but differed from this compound in many properties. He explained, however, every discrepancy with satisfaction to himself; he contrived for himself a theory. Several months later, he received a paper of M. Balard's," announcing the discovery of bromine, " and on that same day he was able to publish the results of experiments on the behaviour of bromine with iron, platinum, and carbon; for Balard's bromine stood in his laboratory, labelled *liquid chloride of iodine.* Since that time he makes no more theories unless they are supported and confirmed by trustworthy experiments; and I can positively assert that he has not fared badly by so doing."

Another point which we notice in the life-work of these two chemists is their untiring labour. They were always at work; wherever they might be, they were ready to notice passing events or natural phenomena, and to draw suggestions from these. As Davy proved the elementary character of iodine and established many of the properties of this substance during a visit to Paris, so we find Dumas making many discoveries during brief visits paid to his friends' laboratories when on excursions away from Paris. During a visit to Aix-les-Bains, he noticed that the walls of the bath-room were covered with small crystals of sulphate of lime. The waters of the bath, he knew, were charged with sulphuretted hydrogen, but they contained no sulphuric acid, nor could that acid be detected in the air of the bath-rooms. This observation was followed up by experiments which proved that a porous material, such as a curtain or an ordinary plastered wall, is able to bring about the union of oxygen with sulphuretted hydrogen, provided moisture be present and a somewhat high temperature be maintained.

Again, we find Liebig and Dumas characterized by great mental honesty. " There is no harm in a man committing mistakes," said Liebig, " but great harm indeed in his committing none, for he is sure not to have worked. . . . An error you have become cognizant of, do not keep in your house from night till morning."

Students of science, more than any other men, ought to be ready to acknowledge and correct the errors into which they fall. It is not difficult for them to do this: they have only to be continually going to Nature; for there they have a court of appeal always ready to hear their case, and to give an absolutely unbiased judgment: they have but to bring their theories and guesses to this judge to have them appraised at their true value.

CHAPTER VII.

MODERN CHEMISTRY.

ON p. 162 I referred to the work of the German chemist Richter, by which the *equivalents* of certain acids and bases were established. Those quantities of various acids which severally neutralized one and the same quantity of a given base, or those quantities of various bases which severally neutralized one and the same quantity of a given acid, were said to be equivalent. These were the quantities capable of performing a certain definite action.

In considering the development of Dumas's substitution theory, we found that Laurent retained this conception of equivalency when he spoke of an equivalent of hydrogen being replaced by an equivalent of chlorine (see p. 272). A certain weight of chlorine was able to take the place and play the part of a certain weight of hydrogen in a compound ; these weights, of hydrogen and chlorine, were therefore equivalent.

This conception has been much used since Laurent's time, but it has for the most part been applied to the atoms of the elements.

Hydrogen being taken as the standard substance, the elements have been divided into groups, in accordance with the number of hydrogen atoms with which one atom of each element is found to combine. Thus certain elements combine with hydrogen only in the proportion of one atom with one atom ; others combine in the proportion of one atom with two atoms of hydrogen ; others in the proportion of one atom with three atoms of hydrogen, and so on.

The adjective *monovalent, divalent, trivalent,* etc., is prefixed to an element to denote that the atom of this element combines with one, or two, or three, etc., atoms of hydrogen to form a compound molecule.

Let us consider what is implied in this statement—"The nitrogen atom is trivalent." This statement, if amplified, would run thus : "One atom of nitrogen combines with three atoms of hydrogen to form a compound molecule." Now, this implies (1) that the atomic weight of nitrogen is known, and (2) that the molecular weight, and the number of nitrogen and hydrogem atoms in the molecule, of a compound of nitrogen and hydrogen are also known.

But before the atomic weight of an element can be determined, it is necessary (as we found on p. 146) to obtain, analyze, and take the specific gravities

of a series of gaseous compounds of that element.
The smallest amount of the element (referred to
hydrogen as unity) in the molecule of any one of
these gases will then be the atomic weight of the
element.

When it is said that "the molecular weight, and
the number of nitrogen and hydrogen atoms in the
molecule, of a compound of nitrogen and hydrogen
are known," the statement implies that the compound in question has been obtained in a pure state,
has been analyzed carefully, has been gasefied, and
that a known volume of the gas has been weighed.
When therefore we say that "the nitrogen atom is
trivalent," we sum up a large amount of knowledge
which has been gained by laborious experiment.

This classification of the elements into groups of
equivalent atoms—which we owe to Frankland,
Williamson, Odling, and especially to Kekulé—has
been of much service especially in advancing the
systematic study of the compounds of carbon. It
helps to render more precise the conception which
has so long been gaining ground of the molecule
as a definite structure.

A monovalent element is regarded as one the
atom of which acts on and is acted on by only one
atom of hydrogen in a molecule; a divalent as
one, the atom of which acts on and is acted on by
two atoms of hydrogen—or other monovalent element—in a molecule; a trivalent element as one,
the atom of which acts on and is acted on by three
atoms of hydrogen—or other monovalent element
—in a molecule; and so on.

The fact that there often exist several compounds of carbon, the molecules of which are composed of the same numbers of the same atoms, finds a partial explanation by the aid of this conception of the elementary atom as a little particle of matter capable of binding to itself a certain limited number of other atoms to form a compound molecule. For if the observed properties of a compound are associated with a certain definite arrangement of the elementary atoms within the molecules of that compound, it would seem that any alteration in this arrangement ought to be accompanied by an alteration in the properties of the compound ; in other words, the existence of more than one compound of the same elements united in the same proportions becomes possible and probable.

I have said that such compounds exist : let me give a few examples.

The alchemists poured a stream of mercury on to molten sulphur, and obtained a black substance, which was changed by heat into a brilliantly red-coloured body. We now know that the black and the red compounds alike contain only mercury and sulphur, and contain these elements united in the same proportions.

Hydrogen, carbon, nitrogen and oxygen unite in certain proportions to produce a mobile, colourless, strongly acid liquid, which acts violently on the skin, causing blisters and producing great pain : if this liquid is allowed to stand for a little time in the air it becomes turbid, begins to boil, gets thicker,

and at last explodes, throwing a white pasty substance about in all directions. This white solid is
inodorous, is scarcely acid to the taste, and does
not affect the skin ; yet it contains the same elements, united in the same proportions, as were present in the strongly acid, limpid liquid from which
it was produced.

Two substances are known each containing
carbon and hydrogen united in the same proportions : one is a gas with strong and irritating odour,
and exerting a most disagreeable action on the eyes ;
the other is a clear, limpid, pleasant-smelling liquid.

Phosphorus is a very poisonous substance : it
readily takes fire in the air at ordinary temperatures,
so that it must be kept under water ; but a modification of phosphorus is known, containing no form
of matter other than phosphorus, which is non-
poisonous, does not take fire easily, and may be
handled with safety.

Once more, there is a compound of nitrogen and
oxygen which presents the appearance of a deep-
red, almost black gas ; there is also a compound
of nitrogen and oxygen which is a clear, colourless
gas ; yet both contain the same elements united
in the same proportions.

But a detailed consideration of *isomerism*, *i.e.*
the existence of more than one compound built up
of the same amounts of the same elements yet
possessing different properties, would lead us too
far from the main path of chemical advance which
we wish to trace.

The chemist is to-day continually seeking to connect the properties of the bodies he studies with the molecular structures of these bodies; the former he can observe, a knowledge of the latter he must gain by reasoning on the results of operations and experiments. His guide—the guide of Lavoisier and his successors—is this: "Similarity of properties is associated with similarity of composition"—by "composition" he generally means molecular composition.

Many facts have been amassed of late years which illustrate the general statement that the properties of bodies are connected with the composition of those bodies. Thus a distinct connection has been traced between the tinctorial power and the molecular composition of certain dye-stuffs; in some cases it has even become possible to predict how a good dye-stuff may be made—to say that, inasmuch as this or that chemical reaction will probably give rise to the production of this or that compound, the atoms in the molecule of which we believe to have a certain arrangement relatively to one another, so this reaction or that will probably produce a dye possessed of strong tinctorial powers.

The compound to the presence of which madder chiefly owes its dyeing powers is called *alizarine;* to determine the nature of the molecular structure of this compound was, for many years, the object of the researches of chemists; at last, thanks especially to the painstaking zeal of two German chemists, it became fairly clear that alizarine and a compound of carbon and hydrogen, called *anthracene*, were closely related in structure. Anthracene was obtained from alizarine, and, after much labour, alizarine was prepared from anthracene. Anthracene is contained in large quantities in the thick pitch which remains when coal-tar is distilled; this pitch was formerly of little or no value, but as soon as the chemical manufacturer found that in this black objectionable mass there lay hidden enormous stores of alizarine, he no longer threw away his coal-tar pitch, but sold it to the alizarine manufacturer for a large sum. Thus it has come to pass that little or no madder is now cultivated; madder-dyeing is now done by means of alizarine made from coal-tar: large tracts of ground, formerly used for growing the madder plant, are thus set free for the growth of wheat and other cereals.

This discovery of a method for preparing alizarine artificially stimulated chemists to make researches into the chemical composition, and if possible to get to know something about the molecular structure of indigo. Those researches have very recently resulted in the knowledge of a series of reactions whereby this highly valuable and costly dye-stuff may be prepared from certain carbon compounds which, like anthracene, are found in coal-tar.

These examples, while illustrating the connection that exists between the composition and the properties of bodies, also illustrate the need there is for giving a scientific chemical training to the

man who is to devote his life to chemical manufactures. Pure and applied science are closely
connected ; he who would succeed well in the latter
must have a competent and a practical knowledge
of the former.

That composition—molecular composition—and
properties are closely related is generally assumed,
almost as an axiom, in chemical researches nowadays.

Lavoisier defined acids as substances containing
oxygen ; Davy regarded an acid as a compound the
properties of which were conditioned by the nature
and by the arrangement of all the elements which
it contained ; Liebig spoke of acids as substances
containing "replaceable" hydrogen ; the student
of the chemistry of the carbon compounds now
recognizes in an organic acid a compound containing hydrogen, but also carbon and oxygen, and he
thinks that the atoms of hydrogen (or some of
these atoms) in the molecule of such a compound
are, in some way, closely related to atoms of oxygen
and less closely to atoms of carbon, within that
molecule,—in other words, the chemist now recognizes that, for carbon compounds at any rate, acids
are acid not only because they contain hydrogen,
but also because that hydrogen is related in a
definite manner within the molecule to other elementary atoms ; he recognizes that the acid or non-
acid properties of a compound are conditioned, not
only by the nature of the elements which together
form that compound, but also by the arrangement

of these elements. Davy's view of the nature of acids is thus confirmed and at the same time rendered more definite by the results of recent researches.

The physical student is content to go no further than the molecule; the properties of bodies which he studies are regarded, for the most part, as depending on the size, the nature, and perhaps the grouping together of molecules. But the chemist seeks to go deeper than this. The molecule is too large a piece of matter for him; the properties which he studies are conceived by him to be principally conditioned by the nature, the number, and the arrangement of the parts of the molecule—of the atoms which together build up the molecule.

In these elementary atoms he has, for the present, found the materials of which the heavens and the earth are made; but facts are being slowly gained which render it probable that these atoms are themselves structures—that they are built up of yet smaller parts, of yet simpler kinds of matter. To gather evidence for or against this supposition, the chemist has been obliged to go from the earth to the heavens, he has been obliged to form a new science, the science of spectroscopic analysis.

This subject has been considered in "The Astronomers," belonging to this series of books; but the point of view from which the matter is there regarded is astronomical rather than chemical. I should like briefly to recall to the reader the fundamental facts of this branch of science.

When a ray of light is allowed to pass through a glass prism and then fall on to a white surface, the image produced on this surface consists of a many-coloured band of light. The blue or violet part of this band is more bent away from the plane of the entering ray than the orange part, and the latter more than the red part of the band. This is roughly represented in Fig. 4, where *r* is the ray of light

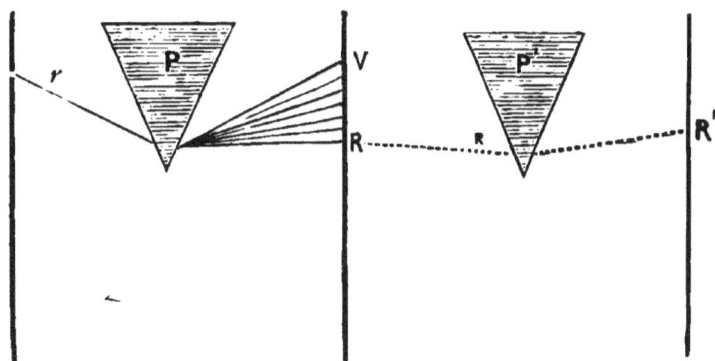

Fig. 4.

passing through the prism P, and emerging as a sevenfold band of coloured lights, of which the violet, V, is most, and the red band, R, is least bent away from the plane of the ray *r*. If the surface—say a white screen—on which the many-coloured band of light, or *spectrum*, falls, is punctured by a small hole, so as to admit the passage of the violet, or blue, or orange, or red light only, and if this violet, etc., light is then passed through a second prism, no further breaking up of that light takes place. This state of matters is represented in the part of the figure towards the right hand, where the red ray, R, is shown as passing through the screen, and falling on to a second prism, P' : the red ray is slightly bent out of its direct course, but is not sub-divided ; it falls on the second screen as a ray of red light, R'.   But if a quantity of the metal sodium is vaporized in a hot non-luminous flame, and if the yellow light thus produced is passed through a prism, a spectrum is obtained consisting of a single yellow line (on a dark background), situated on that part of the screen where the orange-yellow band occurred when the ray of sunlight was split up by the action of the prism.   In Fig. 5 the yellow light from a flame containing sodium is

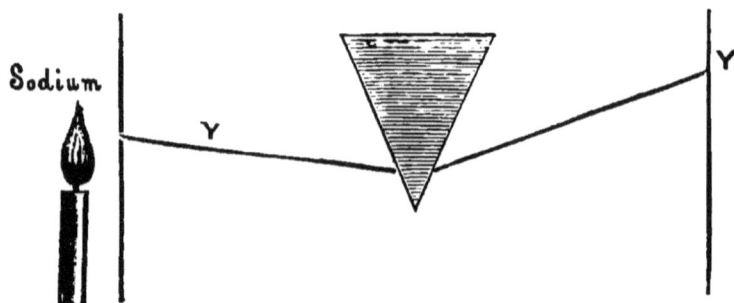

Fig. 5.

represented by the line Y. The light emitted by the glowing sodium vapour is said to be *mono-chromatic.*

Lastly, if the experiment is arranged so that a ray of sunlight or of light from an electric lamp passes through a layer of comparatively cool sodium vapour before reaching the prism, a spectrum is

produced corresponding to the solar spectrum except that a black line appears in the position where
the yellow line, characteristic of sodium, was
noticed in the second experiment.

Fig. 6 represents the result of this experiment:

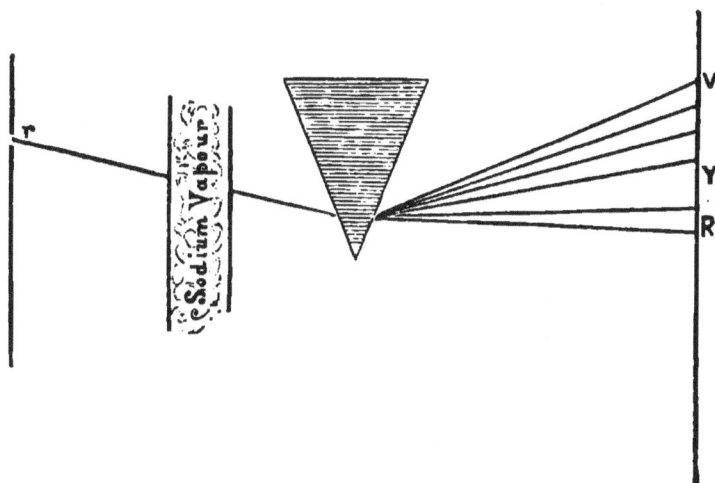

Fig. 6.

the ray of sunlight or electric light, *r*, passes
through a quantity of sodium vapour, and is then
decomposed by the prism ; the spectrum produced
is marked by the absence of light (or by a dark line)
where the yellow line, Y, was before noticed.

These are the fundamental facts of spectroscopic
analysis: sunlight is decomposable into a band of
many colours, that is, into a spectrum ; light
emitted by a glowing vapour is characterized by
the presence of coloured lines, each of which occupies a definite position with reference to the various
parts of the solar spectrum ; sunlight—or the elec-

III.   X

tric light—when allowed to pass through a mass of vapour, furnishes a spectrum characterized by the absence of those bright lines, the presence of which marked the spectrum of the light obtained by strongly heating the vapour through which the sunlight has passed.

The spectrum obtained by decomposing the light emitted by glowing vapour of potassium is characterized by the presence of certain lines—call them A and B lines. We are asked what element (or elements) is present in a certain gas presented to us : we pass a beam of white light through this gas and then through a prism, and we obtain a continuous spectrum (*i.e.* a spectrum of many colours like the solar spectrum) with two dark lines in the same positions as those occupied by the lines A and B. We therefore conclude that the gas in question contains vapour of potassium.

The solar spectrum, when carefully examined, is found to be crossed by a very large number of fine black lines ; the exact positions of many hundreds of these lines have been carefully determined, and, in most cases, they are found to correspond to the positions of various bright lines noticed in the spectra of the lights emitted by hot vapours of various elementary bodies.

Assume that the sun consists, broadly speaking, of an intensely hot and luminous central mass, formed to a large extent of the elementary substances which build up this earth, and that this central mass is surrounded by a cooler (but yet very hot) gaseous envelope of the same elements,—and we have a tolerably satisfactory explanation of the principal phenomena revealed by the spectroscopic study of the sun's light.

On this assumption the central mass of glowing iron, chromium, magnesium, nickel, cobalt, hydrogen, etc., is sending out light; a portion of the light emitted by the glowing iron is quenched as it passes through a cloud of cooler iron vapour outside the central mass, a portion of the light emitted by the glowing chromium is quenched as it passes through a cloud of cooler chromium vapour, and so on; the black lines in the spectrum are the records of these various quenchings of this and that light.

So far then the study of the solar spectrum appears to be tolerably simple, and this study generally confirms the proposition that the material of which the sun is composed is, broadly, identical with those forms of matter which we, on this earth, call the chemical elements.

But whatever be the composition of the sun, it is, I think, evident that in dealing with a ray of light coming therefrom, we are dealing with a very complex phenomenon.

According to the hypothesis which is now guiding us, the solar light which passes into our spectroscope has probably had its beginning in some central part of the sun, and has passed through very thick layers of hot metallic clouds, agitated perhaps by solar cyclones. Could we examine the light coming from some defined part of the sun, we should probably obtain valuable information. During a solar eclipse red prominences are seen projecting beyond the dark shadow of the moon, which covers the sun's disc. Analysis of the light emitted by these prominences has shown that they are phenomena essentially belonging to the sun itself, and that they consist of vast masses of intensely hot, glowing gaseous substances, among which hydrogen is present in large quantities. That these prominences are very hot, hotter than the average temperature of the ordinary solar atmosphere, is proved by the fact that the spectrum of the light coming from them is characterized by bright lines. By special arrangements which need not be discussed here, but which have been partly explained in "The Astronomers" (see pp. 334, 335 of that book), it has been shown that these prominences are in rapid motion : at one moment they shoot up to heights of many thousand miles, at another they recede towards the centre of the sun.

We thus arrive at a picture of the solar atmosphere as consisting of layers of very hot gases, which are continually changing their relative positions and forms ; sometimes ejections of intensely hot, glowing gases occur,—we call these prominences ; sometimes down-rushes of gaseous matter occur,—we call these spots. Among the substances which compose the gaseous layers we recognize hydrogen, iron, magnesium, sodium, nickel, chromium, etc., but we also find substances which can at present be distinguished only by means of the wave-lengths of the light which they emit; thus we have 1474 stuff, 5017 stuff, 5369 stuff, etc.

Let us now turn to another part of this subject. By a special arrangement of apparatus it is possible to observe the spectrum of the light emitted by a glowing vapour, parts of which are hotter than other parts, and to compare the lines in the spectrum of the light coming from the hottest parts with the lines in the spectrum of the light coming from the cooler parts of the vapour. If this is done for sodium vapour, certain lines are apparent in all the spectra, others only in the spectrum of the light coming from the hottest parts of the sodium vapour: the former lines are called "long lines," the latter "short lines." A rough representation of the long and short lines of sodium is given in Fig. 7.

Fig. 7.—Long and short lines of sodium.

Now, suppose that the lines in the spectrum of the light emitted by glowing manganese vapour have been carefully mapped, and classed as long and short lines: suppose that the same thing has been done for the iron lines: now let a little manganese be mixed with much iron, let the mixture be vaporized, and let the light which is emitted be decomposed by the prism of a spectroscope, it will be found that the long lines of manganese alone make their appearance; let a little more manganese be added to the mixture, and now some of the shorter lines due to manganese begin to appear in the spectrum. Hence it has been concluded by Lockyer that if the spectrum of the light emitted by the glowing vapour of any element— call it A—is free from the long lines of any other element—say element B—this second element is not present as an impurity in the specimen of element A which is being examined. Lockyer has applied this conclusion to "purify" various elementary spectra.

The spectrum of element A is carefully mapped, and the lines are divided into long and short lines, according as they are noticed in the spectrum of the light coming from all parts of the glowing vapour of A, or only in the spectrum of the light which comes from the hotter parts of that vapour. The spectra of elements B and C are similarly mapped and classified: then the three spectra are compared; the longest line in the spectrum of B is noted, if this line is found in the spectrum of A, it is marked with a negative sign—this means that so far as the evidence of this line goes B is present as an impurity in A; the next longest B line is searched for in the spectrum of A—if present it also is marked with a negative sign; a similar process of comparison and elimination is conducted with the spectra of A and C. In this way a "purified" spectrum of the light from A is obtained—a spectrum, that is, from which, according to Lockyer, all lines due to the presence of small quantities of B and C as impurities in A have been eliminated.

Fig. 8 is given in order to make this "purify-

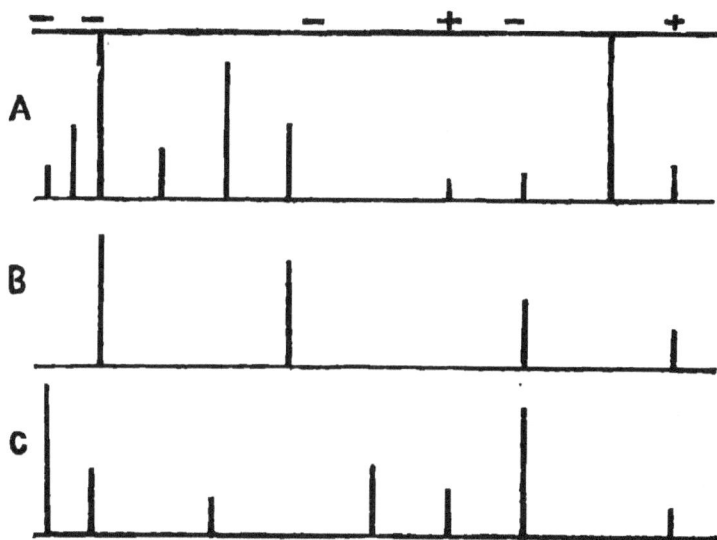

Fig. 8.

ing" process more clearly understood. But when this process has been completed there remain, in many cases, a few short lines common to two or more elementary spectra: such lines are called by Lockyer *basic lines*. He supposes that these lines are due to light emitted by forms of matter simpler than our elements; he thinks that at very high temperatures some of the elements are decomposed, and that the *bases* of these elements are produced and give out light, which light is analyzed by the spectroscope. Such short basic lines are marked in the spectra represented in Fig. 8 with a positive sign.

Now, if the assumption made by Lockyer be admitted, viz. that the short lines, or some of the short lines, which are coincident in the "purified" spectra of various elements, are really due to light emitted by forms of matter into which our so-called elements are decomposed at very high temperatures, it follows that such lines should become more prominent in the spectra of the light emitted by elements the higher the temperature to which these elements are raised. But we know (see p. 308) that the prominences around the sun's disc are hotter than the average temperature of the solar atmosphere; hence the spectrum of the light coming from these prominences ought to be specially rich in "basic" lines: this supposition is confirmed by experiment. Lockyer has also shown that it is the "basic," and not the long lines, which are especially affected in the spectra of light coming from those parts of the solar atmosphere which are subjected to the action of cyclones, *i.e.* which are at abnormally high temperatures. And finally, a very marked analogy has been established between the changes in the spectrum of the light emitted by a compound substance as the temperature is raised, and the substance is gradually decomposed into its elements, and the spectrum of the light emitted by a so-called elementary substance as the temperature of that substance is increased.

But it may be urged that Lockyer's method of
"purifying", a spectrum is not satisfactory; that,
although all the longer lines common to two spectra
are eliminated, the coincident short lines which
remain are due simply to very minute quantities of
one element present as an impurity in the larger
quantity of the other. Further, it has been shown
that several of the so-called "basic" lines are resolved, by spectroscopes of great dispersive power,
into groups of two or more lines, which lines are
not coincident in different spectra.

And moreover it is possible to give a fairly satisfactory explanation of the phenomena of solar
chemistry without the aid of the hypothesis that
our elements are decomposed in the sun into simpler
forms of matter. Nevertheless this hypothesis has
a certain amount of experimental evidence in its
favour; it may be a true hypothesis. I do not think
we are justified at present either in accepting it as
the best guide to further research, or in wholly
rejecting it.

The researches to which this hypothesis has
given rise have certainly thrown much light on the
constitution of the sun and stars, and they have
also been instrumental in forcing new views regarding the nature of the elements on the attention of
chemists, and so of awakening them out of the
slumber into which every class of men is so ready
to fall.

The tale told by the rays of light which travel
to this earth from the sun and stars has not yet

been fully read, but the parts which the chemist has
spelt out seem to say that, although the forms of
matter of which the earth is made are also those
which compose the sun and stars, yet in the sun and
stars some of the earthly elements are decomposed,
and some of the earthly atoms are split into simpler
forms. The tale, I say, told by the rays of light
seems to bear this interpretation, but it is written
in a language strange to the children of this earth,
who can read it as yet but slowly ; for the name
given to the new science was " *Ge-Urania*, because
its production was of earth and heaven. And it
could not taste of death, by reason of its adoption
into immortal palaces ; but it was to know weakness, and reliance, and the shadow of human im-
becility ; and it went with a lame gait ; but in its
going it exceeded all mortal children in grace and
swiftness."

There are certain little particles so minute that
at least sixty millions of them are required to compose the smallest portion of matter which can be
seen by the help of a good microscope. Some of
these particles are vibrating around the edge of an
orb a million times larger than the earth, but at a
distance of about ninety millions of miles away.
The student of science is told to search around the
edge of the orb till he finds these particles, and
having found them, to measure the rates of their
vibrations ; and as an instrument with which to do
this he is given—a glass prism ! But he has accomplished the task ; he has found the minute particles,
and he has measured their vibration-periods.

Chemistry is no longer confined to this earth : the chemist claims the visible universe as his laboratory, and the sunbeams as his servants.

Davy decomposed soda and potash by using the powerful instrument given him by Volta ; but the chemist to-day has thrown the element he is seeking to decompose into a crucible, which is a sun or a star, and awaits the result.

The alchemists were right. There is a philosopher's stone ; but that stone is itself a compound of labour, perseverance, and genius, and the gold which it produces is the gold of true knowledge, which shall never grow dim or fade away.

CHAPTER VIII.

SUMMARY AND CONCLUSION.

WE have thus traced some of the main paths along which Chemistry has advanced since the day when, ceasing to be guided by the dreams of men who toiled with but a single idea in the midst of a world of strange and complex phenomena, she began to recognize that Nature is complex but orderly, and so began to be a branch of true knowledge.

In this review we have, I think, found that the remark made at the beginning of the introductory chapter is, on the whole, a just one. That the views of the alchemists, although sometimes very noble, were "vague and fanciful" is surely borne out by the quotations from their writings given in the first chapter. This period was followed by that wherein the accurate, but necessarily somewhat narrow conception of the Lavoisierian chemistry prevailed. Founded for the most part on the careful, painstaking, and quantitative study of one phenomenon—a very wide and far-reaching phenomenon, it is true—it was impossible that the classification introduced by the father of chemical science should be broad enough to include all the discoveries of those who came after him. But although this classification had of necessity to be revised and recast, the genius of Lavoisier enunciated certain truths which have remained the common possession of every chemical system. By proving that however the forms of matter may be changed the mass remains unaltered, he for the first time made a science of chemistry possible. He defined "element" once for all, and thus swept away the fabric of dreams raised by the alchemists on the visionary foundation of *earth, air, fire* and *water*, or of *mercury, sulphur* and *salt*. By his example, he taught that weighings and measurements must be made before accurate knowledge of chemical reactions can be hoped for; and by his teaching about oxygen being *the acidifier*—although we know that this teaching was erroneous in many details—he showed the possibility of a system of classification of chemical substances being founded on the actually observed properties and composition of those substances.

Lavoisier gained these most important results by concentrating his attention on a few subjects of inquiry. That chemistry might become broad it was necessary that it should first of all become narrower.

The period when the objects of the science were defined and some of its fundamental facts and conceptions were established, was succeeded, as we saw in our sketch, by that in which Dalton departed somewhat from the method of investigation adopted by most masters in science, and by concentrating his great mental powers on facts belonging to one branch of natural knowledge, elaborated a simple but very comprehensive theory, which he applied to explain the facts belonging to another branch of science.

Chemistry was thus endowed with a grand and far-reaching conception, which has been developed and applied by successive generations of investigators : but we must not forget that it was the thorough, detailed work of Black and Lavoisier which made possible the great theory of Dalton.

At the time when Dalton was thinking out his theory of atoms, Davy was advancing as a conqueror through the rich domain which the discovery of Volta had opened to chemistry. Dalton, trained to rely on himself, surrounded from his youth by an atmosphere in which "sweetness and light" did not predominate, thrown on the world at an early age, and obliged to support himself by the drudgery of teaching when he would fain have been engaged in research, and at the same time— if we may judge from his life as recorded by his biographers—without the sustaining presence of such an ideal as could support the emotional part of his nature during this time of struggle,—Dalton,

we found, withdrew in great part from contact with
other scientific workers, and communing only with
himself, developed a theory which, while it showed
him to be one in the chain of thinkers that begins
in Democritus and Leucippus, was nevertheless
stamped with the undeniable marks of his own
individuality and genius, and at the same time was
untouched by any of the hopes or fears, and un-
affected by any of the passions, of our common
humanity.

Davy, on the other hand, was surrounded from
childhood by scenes of great natural beauty and
variety, by contact with which he was incited
to eager desire for knowledge, while at the same
time his emotions remained fresh and sensitive to
outward impressions. Entering on the study of
natural science when there was a pause in the
march of discovery, but a pause presageful of fresh
advances, he found outward circumstances singularly
favourable to his success ; seizing these favourable
circumstances he made rapid advances. Like
Lavoisier, he began his work by proving that there
is no such thing in Nature as transmutation, in the
alchemical meaning of the term ; as Lavoisier had
proved that water is not changed into earth, so did
Davy prove that acid and alkali are not produced
by the action of the electric current on pure water.
We have shortly traced the development of the
electro-chemical theory which Davy raised on the
basis of experiment ; we have seen how facts
obliged him to doubt the accepted view of the

composition of hydrochloric acid and chlorine, and
how by the work he did on these subjects chemists
have been finally convinced that an element is not
a substance which *cannot be*, but a substance which
has not been decomposed, and how from this work
has also arisen the modern theory of acids, bases
and salts.

We found that, by the labours of the great
Swede J. J. Berzelius, the Daltonian theory was
confirmed by a vast series of accurate analyses, and,
in conjunction with a modification of the electro-
chemical theory of Davy, was made the basis of a
system of classification which endeavoured to include all chemical substances within its scope.
The atom was the starting-point of the Berzelian
system, but that chemist viewed the atom as a dual
structure the parts of which held together by reason
of their opposite electrical polarities. Berzelius, we
saw, greatly improved the methods whereby atomic
weights could be determined, and he recognized
the importance of physical generalizations as aids in
finding the atomic weights of chemical substances.

But Berzelius came to believe too implicitly in
his own view of Nature's working ; his theory
became too imperious. Chemists found it easier
to accept than to doubt an interpretation of facts
which was in great part undeniably true, and which
formed a central luminous conception, shedding
light on the whole mass of details which, without it,
seemed confused and without meaning.

If the dualistic stronghold was to be carried, the

attack should be impetuous, and should be led by men, not only of valour, but also of discretion. We found that two champions appeared, and that, aided by others who were scarcely inferior soldiers to themselves, they made the attack, and made it with success.

But when the heat of the battle was over and the bitterness of the strife forgotten, it was found that, although many pinnacles of the dualistic castle had been shattered, the foundation and great part of the walls remained; and, strange to say, the men who led the attack were content that these should remain.

The atom could no longer be regarded as always composed of two parts, but must be looked on rather as one whole, the properties of which are defined by the properties and arrangements of all its parts; but the conception of the atom as a structure, and the assurance that something could be inferred regarding that structure from a knowledge of the reactions and general properties of the whole, remained when Dumas and Liebig had replaced the dualism of Berzelius by the unitary theory of modern chemistry; and these conceptions have remained to the present day, and are now ranked among the leading principles of chemical science; only we now speak of the " molecule " where Berzelius spoke of the " atom."

Along with these advances made by Dumas, Liebig and others in rendering more accurate the general conception of atomic structure, we found

III.   Y

that the recognition of the existence of more than one order of small particles was daily gaining ground in the minds of chemists.

The distinction between what we now call atoms and molecules had been clearly stated by Avogadro in 1811 ; but the times were not ripe. The mental surroundings of the chemists of that age did not allow them fully to appreciate the work of Avogadro. The seed however was sown, and the harvest, although late, was plentiful.

We saw that Dumas accepted, with some hesitation, the distinction drawn by Avogadro, but that failing to carry it to its legitimate conclusion, he did not reap the full benefit of his acceptance of the principle that the smallest particle of a substance which takes part in a physical change divides into smaller particles in those changes which we call chemical.

To Gerhardt and Laurent we owe the full recognition, and acceptance as the foundation of chemical classification, of the atom as a particle of matter distinct from the molecule ; they first distinctly placed the law of Avogadro—" Equal volumes of gases contain equal numbers of molecules "—in its true position as a law, which, resting on physical evidence and dynamical reasoning, is to be accepted by the chemist as the basis of his atomic theory. To the same chemists we are indebted for the formal introduction into chemical science of the conception of types, which, as we found, was developed by Frankland, Kekulé, and others, into the modern doctrine of equivalency of groups of elementary atoms.

We saw that, in the use which he made of the laws of Mitscherlich, and of Dulong and Petit, Berzelius recognized the importance of the aid given by physical methods towards solving the atomic problems of chemistry; but among those who have most thoroughly availed themselves of such aids Graham must always hold a foremost place.

Graham devoted the energies of his life to tracking the movements of atoms and molecules. He proved that gases pass through walls of solid materials, as they pass through spaces already occupied by other gases; and by measuring the rapidities of these movements he showed how it was possible to determine the rate of motion of a particle of gas so minute that a group of a hundred millions of them would be invisible to the unassisted vision. Graham followed the molecules as in their journeyings they came into contact with animal and vegetable membranes; he found that these membranes presented an insuperable barrier to the passage of some molecules, while others passed easily through. He thus arrived at a division of matter into colloidal and crystalloidal. He showed what important applications of this division might be made in practical chemistry, he discussed some of the bearings of this division on the general theory of the molecular constitution of matter, and thus he opened the way which leads into a new territory rich in promise to him who is able to follow the footsteps of its discoverer.

Other investigators have followed on the general lines laid down by Graham ; connections, more or less precise, have been established between chemical and physical properties of various groups of compounds. It has been shown that the boiling points, melting points, expansibilities by heat, amounts of heat evolved during combustion, in some cases tinctorial powers of dye-stuffs, and other physical constants of groups of compounds, vary with variations in the nature, number and arrangements of the atoms in the molecules of these compounds.

But although much good work has been done in this direction, our ignorance far exceeds our knowledge regarding the phenomena which lie on the borderlands between chemistry and physics. It is probably here that chemists look most for fresh discoveries of importance.

As each branch of natural science becomes more subdivided, and as the quantity of facts to be stored in the mind becomes daily more crushing, the student finds an ever-increasing difficulty in passing beyond the range of his own subject, and in gaining a broad view of the relative importance of the facts and the theories which to him appear so essential.

In the days when the foundation of chemistry was laid by Black, Priestley, Lavoisier and Dalton, and when the walls began to be raised by Berzelius and Davy, it was possible for one man to hold in his mental grasp the whole range of subjects which he studied. Even when Liebig and Dumas built the fabric of organic chemistry the mass of facts to be considered was not so overpowering as it is now. But we have in great measure ourselves to blame; we have of late years too much fulfilled Liebig's words, when he said, that for rearing the structure of organic chemistry masters were no longer required —workmen would suffice.

And I think we have sometimes fallen into another error also. Most of the builders of our science—notably Lavoisier and Davy, Liebig and Dumas—were men of wide general culture. Chemistry was for them a branch of natural science; of late years it has too much tended to degenerate into a handicraft. These men had lofty aims; they recognized—Davy perhaps more than any —the nobility of their calling. The laboratory was to them not merely a place where curious mixtures were made and strange substances obtained, or where elegant apparatus was exhibited and carefully prepared specimens were treasured; it was rather the entrance into the temple of Nature, the place where day by day they sought for truth, where, amid much that was unpleasant and much that was necessary mechanical detail, glimpses were sometimes given them of the order, harmony and law which reign throughout the material universe. It was a place where, stopping in the work which to the outsider appeared so

Y 3

dull and even so trivial, they sometimes, listening with attentive ear, might catch the boom of the "mighty waters rolling evermore," and so might return refreshed to work again.

Chemistry was more poetical, more imaginative then than now; but without imagination no great work has been accomplished in science.

When a student of science forgets that the particular branch of natural knowledge which he cultivates is part of a living and growing organism, and attempts to study it merely as a collection of facts, he has already Esau-like sold his birthright for a mess of pottage; for is it not the privilege of the scientific student of Nature always to work in the presence of "something which he can never know to the full, but which he is always going on to know"—to be ever encompassed about by the greatness of the subject which he seeks to know? Does he not recognize that, although some of the greatest minds have made this study the object of their lives, the sum of what is known is yet but as a drop in the ocean? and has he not also been taught that every honest effort made to extend the boundaries of natural knowledge must advance that knowledge a little way?

It is not easy to remember the greatness of the issues which depend on scientific work, when that work is carried on, as it too often is, solely with the desire to gain a formal and definite answer to some question of petty detail.

"That low man seeks a little thing to do,
    Sees it and does it :
This high man, with a great thing to pursue,
    Dies ere he knows it.

"That low man goes on adding one to one,
    His hundred's soon hit :
This high man, aiming at a million,
    Misses a unit."

'INDEX.

G

PRINTED BY WILLIAM CLOWES AND SONS, LIMITED, LONDON AND BECCLES.